T0338574

INTERPRETING FEYERABEND

This collection of new essays interprets and critically evaluates the philosophy of Paul Feyerabend. It offers innovative historical scholarship on Feyerabend's take on topics such as realism, empiricism, mimesis, voluntarism, pluralism, materialism, and the mind-body problem, as well as certain debates in the philosophy of physics. It also considers the ways in which Feyerabend's thought can contribute to contemporary debates in science and public policy, including questions about the nature of scientific methodology, the role of science in society, citizen science, scientism, and the role of expertise in public policy. The volume will provide readers with a comprehensive overview of the topics that Feyerabend engaged with throughout his career, showing both the breadth and the depth of his thought.

KARIM BSCHIR is a Lecturer at the University of St Gallen, Switzerland. His research focuses on topics in the general philosophy of science and the history of philosophy of science. He has written on the relationship of Feyerabend's pluralism to Popper's critical rationalism.

JAMIE SHAW is Postdoctoral Fellow at the Institute for History and Philosophy of Science and Technology, University of Toronto. His primary research interests revolve around the implications of Feyerabend's pluralism for science funding policy.

INTERPRETING FEYERABEND

Critical Essays

EDITED BY

KARIM BSCHIR

University of St. Gallen, Switzerland

JAMIE SHAW

University of Toronto

CAMBRIDGE
UNIVERSITY PRESS

CAMBRIDGE
UNIVERSITY PRESS

University Printing House, Cambridge CB2 8BS, United Kingdom

One Liberty Plaza, 20th Floor, New York, NY 10006, USA

477 Williamstown Road, Port Melbourne, VIC 3207, Australia

314–321, 3rd Floor, Plot 3, Splendor Forum, Jasola District Centre,
New Delhi – 110025, India

79 Anson Road, #06–04/06, Singapore 079906

Cambridge University Press is part of the University of Cambridge.

It furthers the University's mission by disseminating knowledge in the pursuit of
education, learning, and research at the highest international levels of excellence.

www.cambridge.org
Information on this title: www.cambridge.org/9781108471992
DOI: 10.1017/9781108575102

© Karim Bschir and Jamie Shaw 2021

This publication is in copyright. Subject to statutory exception
and to the provisions of relevant collective licensing agreements,
no reproduction of any part may take place without the written
permission of Cambridge University Press.

First published 2021

A catalogue record for this publication is available from the British Library.

ISBN 978-1-108-47199-2 Hardback

Cambridge University Press has no responsibility for the persistence or accuracy of
URLs for external or third-party internet websites referred to in this publication
and does not guarantee that any content on such websites is, or will remain,
accurate or appropriate.

Contents

v

Figures

Tables

Contributors

CHIARA AMBROSIO is Associate Professor in history and philosophy of science at the Department of Science and Technology Studies, University College London. Her research interests include visual culture and the relationship between science and art in the nineteenth and twentieth centuries, scientific representations, American pragmatism, particularly the philosophy of C.S. Peirce, and the legacy of pragmatism in broader debates in history and philosophy of science. Her chapter in this volume is her first contribution to the scholarship on Feyerabend, but she has been a practicing Feyerabendian for most of her academic life.

HAKOB BARSEGHYAN is an Assistant Professor at Victoria College, University of Toronto. He is the author of *The Laws of Scientific Change* (2015), which seeks to uncover patterns of how scientific theories and method change through time. He is also the co-creator of the *Encyclopaedia of Scientonomy* (https://www.scientowiki.com/Main_Page) and the *Journal of Scientonomy* (https://scientojournal.com/index.php /scientonomy).

MATTHEW J. BROWN is the Director of the Center for Values in Medicine, Science, and Technology, Program Head for history and philosophy, and Professor of philosophy and history of ideas at the University of Texas at Dallas. The main areas of his research deal with the intersection of science with values, the way science informs policy, and the history of philosophy of science. His book *Science and Moral Imagination: A New Ideal for Values in Science* (University of Pittsburgh Press) explores the role of values in science and the scientific basis of values from a broadly pragmatist perspective. See more at: http://www.matthewjbrown.net/

KARIM BSCHIR is a Lecturer at the University of St. Gallen, Switzerland. His research focuses on topics in the general philosophy of science and the history of philosophy of science. He has written on the relationship of Feyerabend's pluralism to Popper's critical rationalism.

HASOK CHANG is the Hans Rausing Professor of history and philosophy of science at the University of Cambridge. He received his degrees from Caltech and Stanford, and has taught at University College London. He is the author of *Is Water H₂O? Evidence, Realism and Pluralism* (2012), and *Inventing Temperature: Measurement and Scientific Progress* (2004). He is a co-founder of the Society for Philosophy of Science in Practice (SPSP) and the Committee for Integrated History and Philosophy of Science.

IAN JAMES KIDD is a Lecturer in philosophy at the University of Nottingham. His research interests include the philosophy of science, epistemology, and the history of Austro-German philosophy. He was the editor, with Matthew J. Brown, of *Reappraising Feyerabend* (2016) and is a author of several papers on Feyerabend's philosophy. His website is www.ianjameskidd.weebly.com

DANIEL KUBY is a postdoctoral researcher in the project "Forcing: Conceptual Change in the Foundations of Mathematics" at the University of Konstanz, Germany. His research interests include the history of scientific philosophy and philosophy of science (logical empiricism, Feyerabend); the history and philosophy of modern set theory (forcing); and, more recently, the philosophy of computer science (programming languages).

MARTIN KUSCH is Professor of philosophy of science and epistemology at the University of Vienna. He previously taught in Oulu (Finland), Edinburgh, and Cambridge. His book *Relativism in the Philosophy of Science* is forthcoming from Cambridge University Press, and he recently edited the *Routledge Handbook for the Philosophy of Relativism*.

J.B. MANCHAK is Professor of logic and philosophy of science at the University of California, Irvine. His work centers on the limits of our knowledge concerning the spatio-temporal structure of the universe.

SARAH M. ROE is an Associate Professor of history and philosophy at Southern Connecticut State University. Her research interests include the ethical and social implications of science, generalizations, and explanation within science and the history of medicine.

JAMIE SHAW is a Postdoctoral Fellow at the Institute for History and Philosophy of Science and Technology, University of Toronto. His primary research interests revolve around the implications of Feyerabend's pluralism for science funding policy.

K. BRAD WRAY works at the Centre for Science Studies at Aarhus University in Denmark. His research addresses issues in the social epistemology of science, Kuhn's philosophy of science, and the anti-realism /realism debate in the philosophy of science. His publications include Kuhn's *Evolutionary Social Epistemology* (2011) and *Resisting Scientific Realism* (2018), both with Cambridge University Press. He is also one of the coeditors of the Springer/Nature journal *Metascience*.

Introduction
Paul Feyerabend's Philosophy in the Twenty-First Century
Jamie Shaw and Karim Bschir

This volume aims to develop our understanding of and critically engage with themes in the corpus of Paul Feyerabend. The eleven chapters of the volume collectively pursue two intertwined goals: providing historically robust interpretations of Feyerabend's texts and reinjecting ingredients of Feyerabend's thought into current discussions. Before delving into this, it is worth making some brief remarks on Feyerabend's legacy and the extant secondary scholarship.

Discerning Feyerabend's legacy is a complicated, yet fascinating topic. On the one hand, nearly every philosopher of science has encountered Feyerabend at some point. His texts are a mainstay of philosophy of science courses at all levels; his name is frequently mentioned as a forerunner of the historical turn that reshaped the discipline, and his work is widely discussed outside of academia. In a recent poll, Feyerabend was ranked as the ninth most significant twentieth-century philosopher of science.[1] If one consults Elizabeth Lloyd's (1996, p. 261) list of prominent scholars who have been influenced by Feyerabend, it is understandable that the layout of contemporary philosophy of science reverberates themes from Feyerabend's corpus. 'Some of Feyerabend's claims ... are accepted as *standard* by so many philosophers of science ... they are such a basic part of philosophical background' (pp. 257–258). Looking at things this way, one would think that Feyerabend was markedly influential and should figure prominently in current philosophical discussions concerning science (at least historically). But strangely, this is not the case. The literature that critically engages with his thought in depth is scarce, even on topics that Feyerabend brought to the forefront. For example, in recent volumes on topics which Feyerabend was famous for discussing at length (e.g. scientific method, demarcation, values in science, democracy and science, the

---

[1] https://leiterreports.typepad.com/blog/2010/10/most-significant-philosophers-of-science-of-the-20th-century-the-results.html. Last date accessed: November 18, 2020.

disunity of science), he is barely mentioned, and when he is, it is often in passing. More generally, though, his views are not engaged with either. Amongst his detractors, Feyerabend's thought is often referenced obliquely through the slogan 'anything goes', which stands in for an obviously flawed idea by one of 'the worst enemies of science' (Theocharis and Psimopoulos 1987) that requires no serious attention (see Kusch, this volume). Amongst his sympathisers, Feyerabend's insights are often quickly lumped together with his contemporaries, especially Kuhn's (Bschir 2015). Philip Kitcher admits that 'the awful bogey Kuhn-Feyerabend [was] a chimera constructed by people (including me) who failed to recognize the important differences between these two thinkers' even though 'Feyerabend was very clear that his ideas were different from Kuhn's' (Kitcher 2019, p. 4). While these comparisons are not altogether out of order, their prevalence at the expense of detailing Feyerabend's particular positions has contributed to the lack of engagement with the unique elements of his perspective. Feyerabend's insights have become part and parcel of more general themes in the philosophy of science, rather than those of a unique thinker. With both audiences, Feyerabend's thought has failed to garner detailed attention.

What we are left with is a mixed legacy. Feyerabend is extremely well known and, simultaneously, overlooked. More accurately, most are acquainted with (at least) a coarse-grained depiction of Feyerabend's thought, but the details of his views have escaped attention. As will become clear in this volume, a deeper understanding of Feyerabend may lead to many (perhaps) unexpected fruits.

Feyerabend's early work, from the late 1940s until the late 1950s, mostly focussed on technical issues in quantum mechanics. While these findings were influential on his later thinking (van Strien 2019; Kuby, this volume), they made little impact in physics or philosophy. Feyerabend's work that is more famous emerged in the early 1960s. His landmark 'Explanation, Reduction and Empiricism', published in 1962, became a focal point of philosophical discussion. Here, Feyerabend develops and analyses the notion of 'incommensurability' in the same year as Kuhn in *The Structure of Scientific Revolutions*. Moreover, this paper was filled with in-depth historical case studies. This was becoming fashionable and, since Feyerabend was one of the first to dedicate extensive space for history in his papers, it became noteworthy as an exemplar of historical philosophy of science. Finally, the radical nature of incommensurability led to the re-analysis of many of the basic presumptions of logical empiricism, which was still the primary topic of discussion. Feyerabend's follow-up papers in the mid-1960s, including 'How to Be a Good Empiricist', 'Realism and Instrumentalism', 'Problems of Empiricism, Part I', and 'Reply to Comments', built off Feyerabend's arguments developed in 1962.

After the mark he made in 1962, his philosophical colleagues began to follow Feyerabend's work and, despite frequent and radical disagreements, respected him as a talented philosopher worthy of attention. His papers during the early to mid-1960s were taken seriously and discussed in great detail – as would be expected. There are many remarks from prominent philosophers and physicists who would assent to this statement of Feigl's:

> Immediately, during my first conversation with Feyerabend, I recognized his competence and brilliance. He is, perhaps, the most unorthodox philosopher of science I have ever known. We have often discussed our differences publicly. Although the audiences usually sided with my more conservative views, it may well be that Feyerabend is right, and I am wrong. (Feigl 1968/1981, 668; see also Stadler 2010)

Regardless of how unconventional or perplexing some of Feyerabend's views appeared, his contemporaries felt like his works were worthy of attention. This respect was reciprocal, as Feyerabend thought highly of many of his colleagues, even if he did not always make it explicit. This allowed dialogue to flourish even while disagreeing on some of the most fundamental matters imaginable. It is beyond question that Feyerabend was a highly talented scholar, not a mere contrarian or crank.

Beginning in the late 1960s, the intellectual landscape of philosophy of science changed dramatically. This is largely due to a series of retirements and deaths of many of the most prominent figures in the field. Koyré died in 1964, Hanson in 1967, Carnap in 1970 and Lakatos in 1974. Braithwaite retired in 1966, Popper in 1969, Feigl in 1971, Hempel in 1973, Goodman in 1977, and Polanyi and Quine in 1978. The decreasing plausibility of logical empiricism and an increasingly younger generation of philosophers led to a new philosophical community with a new set of interests.[2] Some of the most popular topics in the early 1980s include confirmation, causation, explanation, feminist perspectives on science, and realism. Feyerabend's work on confirmation had little to do with solving riddles of induction, which he mocked as being unimportant. He wrote next to nothing on causation, and his interest in realism was not related to abductive arguments of Putnam and Boyd (see Chang, this volume). His only work directly relevant to explanation was on incommensurability. It is curious

---

[2] A version of Planck's principle, where intellectual changes correspond with changes in the composition of the community, seems to be operative here (see Zivin et al. 2019).

why Feyerabend's thought is invoked less in feminist philosophy of science, despite being an influence on Longino and Harding. Finally, general philosophy of science became more divided up into philosophy of individual sciences which engaged with general themes concerning science less and less. As such, it makes sense that thinkers writing on these themes would not delve deep into Feyerabend's corpus.

Moreover, the new generation of philosophers was less acquainted with Feyerabend's earlier works and was introduced to his thought mainly through *Against Method* (AM). Despite the fact that AM contained many verbatim passages from his earlier papers, they were blended and surrounded with intense rhetoric and provocative pronouncements like 'anything goes'. This was met with confusion and hostility within the philosophical community. His lectures, which had garnered a reputation within and outside of academia (see below), were provocative and packed full of hyperbole, jokes and riddles. At one point, even Lakatos, upon hearing Feyerabend claim that Aristotelian mechanics had greater empirical content than classical mechanics, screeched 'Oh Paul! How can you say such a thing?' (quoted in Gillies 2019, p. 107). This contributed to a newfound perception of Feyerabend as a provocateur rather than a dignified philosopher in the search for truth. The respect Feyerabend earned from his previous colleagues was absent, and his new techniques of presentation led to his gradual irreverence from the perspective of many of his new interlocutors.

Another major source of contention was that Feyerabend frequently spoke favourably about systems of thought, such as Aristotelian physics or astrology, which, from the perspective of modern science, were seen by many as either outdated or simply false. What many of his contemporaries did not understand was that Feyerabend did not defend these systems, because he thought that they were true in the strict sense. His interest in unconventional theories and world views arose mainly from the fact that they created meaning and value for those who held them. For Feyerabend, it was this feature that made them worth studying. Similarly, he attacked rationalism and a purely scientific outlook on life and reality neither because he was against science in principle nor because he thought that rationality is intrinsically bad. His attacks were mainly directed against those philosophers or scientists who propagated rationalism and science with hegemonic ambitions (see Shaw 2020). His proclivity to react to criticism with even more rhetoric and provocation certainly did not help much to clarify existing misunderstandings.

This is on display when one analyses the reviews of AM. The reception of AM was overwhelmingly negative, with few exceptions (e.g. Naess 1975).

Feyerabend's views were characterised as 'absurd', 'not worth taking seriously', 'between trivial and false', and 'unphilosophical'. He was not given
the benefit of the doubt which followed from the intellectual respect of his
previous colleagues; the (mostly) new generation who were not directly
aware of Feyerabend's prowess saw little reason to engage with his texts
deeply. Most of the strawmen that have become associated with
Feyerabend's views came from these reviews and persist until today (see
Shaw 2017, pp. 7–8). He also admits that laziness played a role in his
replies, such that he did not try in earnest to make himself understood
(Feyerabend 1995, pp. 144–145). Moreover, Feyerabend's response to this
tirade was less than productive (see Kidd, this volume). He was angry and
depressed, leading him to call his critics 'illiterates' and question their
ability to understand basic argumentative norms. He even conjectured
why this response came about, noting the youth of his reviewers.[3] Given
how widespread and common the 'mistakes' of their reviews were,
Feyerabend hypothesised that AM was received poorly simply because
philosophers only knew philosophy and were unfamiliar with the rhetorical techniques he used; they 'have read only Popper and Carnap and who
have never heard of Heine, Mencken or Tucholsky' (Feyerabend
1978a, p. 38).

As a consequence, Feyerabend had been feeling increasingly isolated in
academia. His best philosophical adversary and friend, Lakatos, had suddenly and unexpectedly died in 1974. It makes sense, given this, that
Feyerabend would turn his back on the philosophical community as
a result of the less-than-ideal feedback he felt he received. After these
vitriolic retorts, Feyerabend published less on the themes of his earlier
work, focusing more on politics (see Brown, this volume), history, anthro-
pology and art (see Ambrosio, this volume). He gradually removed himself
from the academic community, even stating that he was 'not a philosopher'
(Feyerabend 1993) and arguing that philosophy of science should be
defunded in his later life (Feyerabend 1994a). The fact that Feyerabend
never offered a level-headed defence of his earlier views further led to his
ostracisation within the community.

Finally, an additional contributing factor to a decline in interest in
Feyerabend's work concerns his students. A well-known mechanism to

---

[3] In Feyerabend's own words: 'Against Method (AM) is my first book and the first work whose reviews
I studied in some detail. In the course of this study, I discovered two things. Most reviewers are
'young' people whose careers started one or two academic generations after the Kuhn-Lakatos era;
and their reviews (with some rare exceptions here and there) have certain features in common'
(Feyerabend 1978a, p. 37).

keep interest in particular topics alive is to supervise successful students who continue to engage with these topics. Despite Lakatos' sudden death, many of his views continued to be discussed, even if in an insular fashion, given the professional successes of his doctoral supervisees who wrote extensively on his philosophical views. Most of Feyerabend's students did not publish a great deal on his work and focussed on other topics (by and large).[4] Given Feyerabend's personality, it makes sense that he wouldn't feel inclined to direct his students towards his own work. In a letter to Lakatos, Feyerabend writes that

> I like to hear that I am an important Figure, but I also *don't* like to hear it . . . Because then think I should do this and I should do that, write and write, make things clear, correct here, and correct there, make things 'perfectly clear' where I have been misunderstood, and for the like of me I do not want to be dragged down into the sewers by tendencies of this kind. I want to live a quiet, peaceful, lazy life. (in Motterlini 1999, p. 281)

While there is certainly nothing problematic here, it had the consequence of further contributing to a lack of serious engagement with his views.

Outside of philosophy, though, Feyerabend made a more positive impression. As Preston observes, 'the real impact he made was outside philosophy' (Preston 2017). *Against Method* was an international best seller and the sentiment often voiced was that it was a liberating text that empowered younger scientists to reject the teachings of previous generations and experiment with new approaches. This message especially resonated within the social sciences, who were under intense pressure to conform to 'the standards' of the natural sciences (see Solovey 2013). Moreover, Feyerabend himself became an increasingly popular personality outside academia and his lectures became legendary and were massively attended. One participant writes that '[t]he lecture hall was always packed with eager students who even sat on the floor in front of the podium. He was very entertaining and fun to listen to' (Meyer 2011). His sphere of influence stretched outside academia, so the lack of engagement with Feyerabend in print should not lead us to believe that he did not leave a mark.[5]

Feyerabend's thought was not entirely overlooked in print either. Most who studied him closely and sympathetically, though, did not work in philosophy. His texts became increasingly prominent within the newly

---

[4] An incomplete list of Feyerabend's doctoral students include Anna Carolina Krebs Pereira Regner, Sheldon Reaven, Maurice Finocchiaro, Gonzalo Munévar, Nancey Murphy and Neil Thomason.
[5] For a current example, see Smolin (2006).

burgeoning field of science and technology studies. Many of the founders of the field claimed that Feyerabend was an important influence on their views, and their naturalistic depictions of science were developed with a Feyerabendian spirit. However, these scholars were ideologically and institutionally segregated from the philosophical community. The 'Science Wars' largely took disciplinary lines, with the science study scholars on one side and philosophers on the other (Brown 2001; Sismondo 2005). Feyerabend's work, in both communities, was seen to fall within the former camp despite his rank as a philosopher. He was lumped together with many others who denied the rationality or objectivity of science (e.g. in Sokal and Bricmont 1998). Again, we see Feyerabend's work being taken as a token of a larger movement. As a result, Feyerabend's work became increasingly prominent outside philosophy and increasingly dwindling within philosophy.

Against this grain, in the past thirty years, there has been a staggered increase of interest in Feyerabend's thought. In the late 1980s, George Couvalis published a series of articles and a book, *Feyerabend's Critique of Foundationalism*, on Feyerabend. Two of the papers focussed on Feyerabend's philosophy of drama and another on Feyerabend's case study on Brownian Motion. Independently, a collected volume, *Beyond Reason* (edited by Gonzalo Munévar) was released in 1991. This volume contained an impressive selection of papers on a wide variety of topics from notable scholars in the field. In the mid to late 90s, a pair of influential papers by Elizabeth Lloyd were published and a series of papers and a book, *Feyerabend: Philosophy, Science and Society*, were composed by John Preston. From here, the literature on Feyerabend gained steam. A second collected volume, *The Worst Enemy of Science?* (edited by D. Lamb, J. Preston and G. Munévar), was released in 2000 followed by a series of papers and books by Robert Farrell and Eric Oberheim. Since then, there has been an exponential increase of interest in Feyerabend's philosophy, with numerous conferences and papers devoted to Feyerabend including important contributions by authors in this volume. It seems like now, more than ever, we are in a position to start taking Feyerabend seriously again.

This literature has mostly focussed on understanding Feyerabend's perspective rather than critically engaging it or applying it to contemporary topics. This is not to say that there is a lack of criticism altogether. Preston, for example, is vocal about his grievances with Feyerabend's conception of semantics and scientific method. Moreover, there are excellent examples of papers that engage Feyerabend's thought with contemporary issues (e.g. Brown 2009; Wray 2015; Sorgner 2016; Stuart 2020). We hope that this volume will amplify this undercurrent and shine a spotlight on Feyerabend's corpus to see what insights have yet to be unearthed. Of course, we are not claiming that this lack of criticism is due to the fact that there is none to be had. Rather, there is no proof (in publication, at least) that Feyerabend's views have been found to be fallacious and this justifies their absence. It is this point that we hope to rectify in this volume by showing the continuing fecundity of Feyerabend's corpus.

The eleven chapters in this volume nicely reflect Feyerabend's intellectual breath and diversity. The first six chapters (Ambrosio, Chang, Barseghyan, Wray, Kusch, Shaw) discuss various central themes within Feyerabend's philosophy. Chiara Ambrosio's chapter engages with Feyerabend's writings on art, which have received little attention compared to his works in philosophy of science. Focussing on Feyerabend's treatment of representation across art and science, and in particular on his treatment of the role of imitation in representative practices, Ambrosio shows that art provided Feyerabend with a powerful metaphor to rethink imitation as a relational and performative process, in line with the very development of his later philosophy.

Hasok Chang explores the evolution of Feyerabend's pluralistic realism. Contrary to the common assumption that Feyerabend must have discarded realism as he became more flamboyantly pluralist in his later years, Chang demonstrates that Feyerabend made realism even stronger by adding a metaphysical layer of pluralist realism to the longstanding epistemological layer thereby resolving some tensions in his earlier views, which are revealed in his mid-career attempt to distinguish different meanings of 'realism'.

Drawing on Feyerabend's particularist axiology and his lack of appreciation for key historiographic distinctions such as between theory acceptance and pursuit or between explicit and implicit assumptions of epistemic agents, Hakob Barseghyan's shows why Feyerabend was never able to articulate a general theory of scientific change.

K. Brad Wray's chapter puts Feyerabend's defence of theoretical pluralism in contrast with contemporary defences of pluralism. Wray argues that in his defence of theoretical pluralism, Feyerabend emphasised the importance of comparative evaluation in determining the epistemic merits of competing theories. Feyerabend's rationale for defending pluralism was never to highlight the fact that, when scientists consider more theories, they are more likely to hit the true theory.

Martin Kusch compares and contrasts Feyerabend's best known work AM with the central ideas of Bas van Fraassen's *The Empirical Stance*, one of the few contemporary philosophers of science who engage closely and charitably with Feyerabend's work. Kush draws an insightful parallel between Feyerabend's epistemological anarchism and van Fraassen's epistemic voluntarism and his notion of philosophical 'stances'. Kusch's goal is to arrive at a better understanding of anarchism and epistemic relativism in the philosophy of science.

Jamie Shaw draws attention to Feyerabend's philosophy of mind and to his contributions to the mind–body problem. Contrary to common interpretations, Shaw argues that Feyerabend should not be seen as an early proponent of materialism. Rather, Feyerabend believed that the mind–body problem admits many different solutions, which are to be sorted out as science progresses. Shaw also shows how Feyerabend's view evolves from a methodological to an ethical view on what a proper solution to the mind–body problem would entail.

Chapters 6 and 7 analyse the relevance of Feyerabend in the philosophy of physics. While one takes a historical exegetic approach (Kuby), the other provides an application of Feyerabend's ideas to current discussions in the philosophy of physics (Manchak). Daniel Kuby's contribution turns to Feyerabend's earlier philosophy of physics, which typically has not received the amount of attention as his later considerations in the general philosophy of science have. Kuby offers a specific interpretation of how Feyerabend came from a Popperian critique of the Copenhagen interpretation to a detailed re-evaluation of Niels Bohr's idea of complementarity. Kuby also argues that Feyerabend's intellectual development in physics provides the backdrop for his thoroughgoing turn from methodological monism to methodological pluralism, for which he would have been known to a wider audience.

In a perfectly Feyerabendian fashion, J.B. Manchak ventures into a 'joyful experiment' in the philosophy of physics by exploring 'unreasonable' models of the universe within the general theory of relativity. Manchak's experiment leads to two insightful ideas: first, a stimulating analysis about the potential significance of seemingly unreasonable models, and, second, a critical reflection on the distinction between reasonable and unreasonable ideas in science more generally.

The last three chapters (Kidd, Brown, Roe) draw attention to some of Feyerabend's ideas on the role of science in society that emerge in his later philosophy. Ian James Kidd engages in a thorough analysis of Feyerabend's frequently discussed critique of scientism thereby clarifying a central misunderstanding in many interpretations, namely that Feyerabend's critique of scientific reason must not be taken as a critique of science per se, but rather as a critique of scientism and misplaced estimation of the power and value of science in human life.

The focus of Matthew J. Brown's chapter is Feyerabend's *Science in a Free Society* (1978b). While it remains one of Feyerabend's least well-regarded works and undoubtedly contains many controversial claims, Brown argues that the book is in many aspects ahead of its time, asking normative questions about the place of science in democracy that have only come into fashion more recently.

Sarah Roe applies Feyerabend's remarks on scientific expertise and the relations between science and the public to current discussion about citizen science. Roe shows how Feyerabend's insights may offer us a better understanding of how citizen science can best promote scientific education, offer broader knowledge to participants, increase citizen interest in conservation and policy, increase both citizen local and national engagement and promote a rewarding experience for both the expert and citizen.

We hope this volume constitutes the first set of steps in revitalising an interest in Feyerabend's thought. The range of topics he covered is seemingly endless and the depth of his thought will require great efforts to uncover. To a large extent, philosophy of science marched forward without taking a close look at his perspective. In this volume, we will get a sense of how unfortunate this historical unfolding was.

We would like to thank Cambridge University Press for the opportunity to edit this volume. Matteo Collodel and Paul Hoyningen-Huene have provided support and advice through the process. Hakob Barseghyan has also been of a great help in formatting the volume. Last but not the least, we thank all the authors for their contributions and their patience. We hope that this volume will inspire a great interest in Feyerabend's work.

CHAPTER 1

Feyerabend on Art and Science

Chiara Ambrosio

1.1 Introduction

Paul Feyerabend's philosophy is replete with artistic metaphors. From theatre to literature, music and painting, the arts were used by Feyerabend not merely as decorative examples to showcase a form of contrived erudition, but as a coherent conceptual framework to articulate key methodological and epistemological questions. With a few isolated exceptions (Couvalis, 1987; Brown, 2009; Kidd, unpublished manuscript), philosophers of science have paid little attention to this intriguing and extremely fruitful aspect of Feyerabend's work.

In this chapter, I bring together several strands of Feyerabend's history and philosophy of art and place them in dialogue with the pluralist outlook that characterises his philosophy of science. Scholars have recently re-evaluated Feyerabend's pluralism as a positive thesis running as a coherent thread throughout the various developments of his thought (Preston 1997a; Shaw 2017; and to a certain extent Oberheim 2006 – insofar as he sees pluralism as Feyerabend's response to and attack on conceptual conservativism). Art was part and parcel of this philosophical and pluralist strategy. It is in the background of Feyerabend's early critique of empiricist accounts of observation and experience (Feyerabend 1962, 1965); it is the springboard to launch into a celebration of styles, to demonstrate the dynamic character of early philosophies of nature and their functioning as coherent worldviews (Feyerabend 2016); it is the foil against which arguments about incommensurability and critiques of progress could be tried and tested (Feyerabend 1975a, 1984); it affords a concrete opportunity for blurring the lines between theory and practice (Feyerabend 1994, 1996), in a way that resonates with analogous debates in the historiographies of science and art alike (Hacking 1983; Shapin 1989; Smith 2004; Field 2004, 2016).

Feyerabend's views on art and choices of examples from artistic practice are as varied as the arguments they are intended to support. Here I will

concentrate on his views on representation – a particular line of investigation, which seems to emerge as a recurrent motif especially in his discussions
of the visual arts in relation to science. I will start from his late writings,
where issues of representation are central to his return to the 'problem of
reality' (Feyerabend 1999; see also Kidd 2010).[1] This is perhaps the aspect of
Feyerabend's posthumously published book, *Conquest of Abundance* (1999),
which has been connected more explicitly with his writings on art
(Oberheim 2006, p. 23; Brown 2009, pp. 216–217). But I also want to
show that the discussion of art – particularly of projective techniques in
the invention of perspective – in *Conquest of Abundance* is the culminating
point of a much longer journey, which saw Feyerabend wrestling with the
'naïvely imitative philosophies' lurking in the background of empiricist as
well as realist positions in philosophy of science.

A turning point in this journey is the book *Science as Art* (1984),[2] which
introduces examples and arguments that Feyerabend would revisit over
a decade later, in *Conquest of Abundance*. Comparing these two texts, I will
single out two interconnected lines of inquiry that characterise
Feyerabend's approach to representation. One is the pervasiveness of the
issue of artistic styles, which Feyerabend exploits as a springboard to
question 'naïvely imitative' views in science. In this, I argue, he adopts
a distinctive art historical methodology, which can be traced back to the
anti-mimetic legacy of the Vienna School of Art History.[3] The other is the
question of imitation as a conceptual category in its own right, and its
relation to representation. Here Feyerabend's ideas shift and align with the
various phases of his philosophy. *Science as Art*, written in the mid-1980s,
draws on the analogy between styles in art and science to expose the flaws
inherent in a linear notion of scientific progress. In this context, 'naïvely
imitative philosophies' form the core of Feyerabend's attack against
a narrow conception of progress construed as increasing fidelity to nature.
Conquest of Abundance, on the other hand, rescues a role for imitation as
a dynamic and performative category, which can be productively carried
over from the arts to science, and which is in tune with the exploration of

[1] I follow the periodisation of Feyerabend's philosophy proposed by Brown and Kidd (2016, p. 3).
[2] All my references to *Science as Art* are from the Italian edition, in the Bibliography as Feyerabend
(1984).
[3] The Vienna School has a long and fascinating history, which has recently been revisited by art
historians. See, for instance, Rampley (2013) and Elsner (2006). I will focus on two figures in
particular, Alois Riegl and Sir Ernst H. Gombrich, who were direct influences on Feyerabend.
Gombrich in particular might not be recognised as the most representative member of the Vienna
School, but his recurrent criticism of Riegl and the acknowledged influence of Riegl on his *The Sense
of Order* (1979) justify inscribing him at least in the School's critical legacy.

'the richness of Being' distinctive of the late Feyerabend. If there is a space for imitation in science (and for the late Feyerabend this is indeed the case, though he never reduced representing to imitating), I argue, it is precisely in this performative sense, as an invitation to explore how reality is reconfigured in the process of imitating it.

1.2 'The Ugly Madonna of Siena'

Chapter 4 of Feyerabend's *Conquest of Abundance* opens with an intriguing discussion of the 'ugly Madonna of Siena' (Feyerabend 1999, p. 89), the so-called *Madonna dagli Occhi Grossi* (Figure 1.1). The painting, produced in the second half of the thirteenth century and attributed to the Maestro di Tressa, occupied the high altar of the Duomo of Siena and was believed to have protected the Sienese army against the Florentine invaders at the battle of Montaperti in 1260 (Emmerson 2013, p. 180). The painting's name ('Madonna with big eyes') does not refer to the image itself, but to the eye-shaped ex-voto that surrounded it. As Feyerabend remarks, the image 'worked miracles' (Feyerabend 1999, p. 89), especially in its ability to 'mediate spiritual powers' (Feyerabend 1999, p. 92). Miraculous capacities notwithstanding, the *Madonna dagli Occhi Grossi* was soon found to be inadequate to the prominent place it occupied and was replaced 'by a suitable altarpiece of equal grandeur [as the altar]': Duccio di Buoninsegna's *Maestá* (1308–1311) (Emmerson 2013, p. 180).

Indeed, when contrasted with later images, the painting may be judged as hopelessly unrefined: perched on a backless throne, the Madonna lacks depth, roundness and perspective. Her arms are far too short and hold rather unnaturally the child in her lap. Her somewhat baffled expression appears more like an accident of the painting process than an intentional artistic choice. Feyerabend proceeds to compare the 'ugly Madonna of Siena' with an image produced a quarter of a century later, Raphael's *Madonna del Granduca* (Figure 1.2). Drawing on an old trope in the history of art, he shows that the latter image could easily be judged as an 'improvement' on the clumsy style of the former. This was the common-sense view of artistic representation that historians of art inherited from Giorgio Vasari's *Lives of the Artists* (c. 1550), where the trajectory from artworks like the *Madonna dagli Occhi Grossi* to the *Madonna del Granduca* is described as one of progress towards an increased fidelity to nature. Of Raphael, for example, in a passage cited by Feyerabend himself, Vasari states: 'His figures expressed perfectly the character of those they represented, the modest or the bold being presented just as they are. The children in his pictures were depicted now with mischief in

Figure 1.1 Maestro di Tressa, *Madonna dagli Occhi Grossi* (c. 1225).
© Opera della Metropolitana ONLUS, Aut. N. 207/2020

Figure 1.2 Raphael, *Madonna del Granduca* (c. 1505). Palazzo Pitti, Florence.
© Le Gallerie Degli Uffizi

their eyes, now in playful attitudes. And his draperies are neither too simple nor too involved, but appear wholly realistic' (Vasari [1550] 1979, p. 252).

The old trope of art progressing towards an increasing fidelity to nature is especially attractive to Feyerabend. Elaborating on Vasari, he reconstructs it as 'the imitative view' of art: 'Artists, says Vasari, try to represent real things and events. They do not immediately succeed; held back by ignorance and false traditions they produce stiff and crude images of lamentable proportions. But they gradually improve' (Feyerabend 1999, p. 90). This narrow view of representational success as increasing mimetic conformity to nature hardly constituted the canon in history of art in Feyerabend's time.[4] But a critique of the legacy of Vasari's ideas, and more broadly of Renaissance art as the pinnacle of naturalistic representation, had been especially important in the establishment of history of art as a discipline in its own right.[5] What Feyerabend found particularly congenial in this strand of historical literature was the critique, conducted on empirical as well as theoretical grounds, of the marriage of progress and increased fidelity to nature. Vestiges of a similarly naïve imitative philosophy, he noted, still lurked in the background of contemporary celebrations of 'the unprejudiced scientist who avoids speculation and "tells it like it is"' (Feyerabend 1999, pp. 91–92). The very idea of artists 'gradually improving' towards more realistic representations had a counterpart in both naïve empiricist and naïve realist accounts of science, which had formed the target of Feyerabend's philosophy all along. What made these positions naïve was an implicit, and narrow, form of representationalism, which Feyerabend aimed to expose through his comparison with art: representationalism about sense data as the immediate, uniform and stable contents of observation in the case of empiricism, and representationalism about the coherent, stable and

---

[4] The dawn of mimetic accounts of art and of a conception of representational success as increasing fidelity to nature is traditionally associated to the rise of the artistic avant-gardes. Arthur Danto (1986; 1997), for instance, famously argued that the very notion of progress in the arts began to falter with the concomitant faltering of mimesis as a criterion for artistic representation from the second half of the nineteenth century onwards. Danto's reconstruction is by no means uncontentious; see, for example, Halliwell's (2002, pp. 369–370) criticism of the overly uniform view of mimesis implicitly built in his account.

[5] As I will discuss later, the two sources most cited and used by Feyerabend, Riegl and Gombrich, were both strong opponents of the mimetic tradition and of the very idea of progress towards increasing naturalistic fidelity in the arts. See Riegl [1893] 1992, Riegl [1901] 1985 and Gombrich [1960] 2002. For an overview of the concept of style in relation to ideas of progress in the arts from Vasari to Feyerabend, see Ginzburg (1998).

unified structure of the world which successful scientific theories aim to mirror, in the case of scientific realism.[6] Drawing on art for Feyerabend fulfilled a methodological aim with a clear epistemic import: to show, through visual as well as verbal arguments, the shortcomings of such narrow philosophical accounts of science.

It is important to note (and I will return to this point later) that the targets of Feyerabend's criticism here are neither representationalism nor imitation per se. In *Conquest of Abundance* he does in fact acknowledge that '*there are* artists who want to copy nature, and some succeed to a surprising degree' (Feyerabend 1999, p. 93). The question, as it is often the case in Feyerabend's writings, is how to reconcile imitation as one possible aim of representation with the inherent pluralism of artistic (and by implication, scientific) styles, even when they purport to copy faithfully from nature. The *Madonna dagli Occhi Grossi*, Feyerabend points out, '*may* have caught an element of reality that had disappeared by the time of Raphael – but this must be determined by research, not by metaphysical speculations about "the nature of reality"' (Feyerabend 1999, pp. 93–94). Thus questioning a common-sense view of imitation in *Conquest of Abundance* aims to pave the way for a richer account of what is more broadly entailed in the *process of representing*, by showing the inherent complexity of even the most straightforward cases of artists directly 'copying from nature'. This is an issue that Feyerabend had started exploring much earlier in his writings, and to which I turn in the next section.

1.3   Empiricism and Naïve Representationalism

Conquest of Abundance is neither the first nor the sole text in which art appears as part of Feyerabend's argumentation. As early as 1965, in the essay 'Problems of Empiricism', Feyerabend indulges in a long footnote,[7] complete with images, to advance a historicised and contextual account of observation in response to the dominant empiricist view. His target there is the uniform and stable account of observation

---

[6] Matt Brown (2016) has characterised this narrow and monistic version of realism as 'scientific materialism', and opposed it to Feyerabend's (late) 'abundant realism'. I will return to Brown's position later on, as it offers a metaphysical counterpart to the reformulation of *mimesis* I pursue in this chapter. For a detailed account of Feyerabend's views on realism and their compatibility with his pluralism, see also Hasok Chang's account in chapter 2 of this volume .

[7] I am especially grateful to Matteo Collodel for alerting me to the existence of this rather precious footnote in Feyerabend's corpus.

implicit in the empiricist theses that ideas derive from sensory experience (aided or unaided by instruments) and that the truth of statements containing ideas thus formed can be straightforwardly verified by observation (Feyerabend 1965, p. 147).[8] Lurking in the background of Feyerabend's criticism is a specific concern about the status of observational reports and their treatment in empiricist accounts of science. As Feyerabend scholars have pointed out (Oberheim 2006; Kuby 2015), this concern is directly related to his critique of phenomenalist sense-data epistemologies – the idea that sense data are the immediate objects of perception, and that statements about sense data enjoy a certainty that other kinds of statements lack. It is also a criticism of the ways in which some logical empiricists tried to avoid the identification with phenomenalist positions (e.g. Hempel 1952) by arguing that observational statements report directly observable and intersubjectively testable facts about physical objects. Both these variants of 'radical empiricism', according to Feyerabend, revolved around 'the common belief that experience contains a factual core that is independent of theories' (Feyerabend 1965, p. 151), which ultimately fixed the meaning of observation statements. 'Problems of Empiricism' argues against the idea of a factual or 'given' core and advances instead the claim that sensations and perceptions are at best *indicators* that function in a manner similar to physical instruments. This is also known as Feyerabend's version of the 'pragmatic theory of observation':[9] sensations and perceptions indicate that something exist, but they become descriptions of what exists only when used in a theory which provides their interpretation (Feyerabend 1962, pp. 36–37; Feyerabend 1965, pp. 214ff).

The reference to art features in the very opening of 'Problems of Empiricism' and supports Feyerabend's general point that, for a start, what counts as an 'observational report' has been contentious across

---

[8] Feyerabend's critique of empiricism in the essay is admittedly much broader, and it is in line with the features that Brown and Kidd (2016) have identified as distinctive of his early philosophy: a defence of theoretical pluralism within science, as opposed to the monistic view implicitly built in 'radical' forms of empiricism; a critique of verificationism and phenomenalist sense-data epistemologies; and a commitment to a variety of semantic or conjectural realism in interpreting scientific theories. For reasons of space, I can only address some of these aspects of Feyerabend's early philosophy briefly, and I have chosen parts of his criticism of empiricism that are more explicitly in dialogue with his treatment of art in footnote 8 of 'Problems of Empiricism'.

[9] For detailed discussions of Feyerabend's pragmatic theory of observation, see Kuby (2015), and especially Kuby (2018) which reconstructs in detail the relationship between Feyerabend and Carnap's respective versions of the theory.

history. It also sets up the stage, through a psychological argument, for his criticism of a 'given' core in experience, and particularly in the process of observation. It is here that a long footnote takes him into a detour on the dependence of perception upon belief, and from there to art:

> That primitive people . . . live in an observational world very different from our own is shown by their art. It has been assumed for some time, no doubt under the influence of empiricism, that the 'primitive' character of these productions is due to lack of skill: these people live in the same perceptual world as we do, but they are unable to produce adequate copies of it. (Feyerabend 1965, p. 221 fn. 8)

The 1965 version of Feyerabend's argument runs along similar lines as the discussion of the *Madonna dagli Occhi Grossi* in *Conquest of Abundance*: naïvely imitative philosophies assume that there is a single, unified and stable perceptual world, and that it is the artist's (or scientist's) task to produce an adequate copy of it. But this representational realism, Feyerabend continues, is an 'impossible doctrine':

> It assumes that there is only one correct way of translating occurrences in the three dimensional real world into situations portrayed in an altogether different medium. The world is as it is. The picture is not the world. What then, does the realist demand? He demands that the conventions to which *he is accustomed* (and which are only a meagre selection from a much wider domain of conventions) be adopted. That is, he makes *himself* the measure of the reality of things – the very opposite of what the realistic *doctrine* would allow. (Feyerabend 1965, p. 221 fn. 8)

At this point, Feyerabend's footnote explicitly turns to a classic study on the relation between perception and pictorial conventions in the arts: Sir Ernst Gombrich's *Art and Illusion* ([1960] 2002). Starting precisely from the legacy of Vasari's idea of progress in history of art, Gombrich argued that pictorial realism involved much more than just faithfully copying from an art-independent reality. Instead, convincing figurative representations are *illusions*, which involve the manipulation of inherited perceptual 'schemata' that designate reality by convention. It is the totality of these conventions at a particular time in history, according to Gombrich, that defines a pictorial *style* (Gombrich [1960] 2002, p. 246). The history of art, in his account, consisted in a sustained empirical and theoretical investigation precisely into the dynamics that underpin the rise and fall of pictorial styles, which he also saw as the basis of artistic change and of the inherent pluralism that characterised artistic representations across history.

Gombrich famously built his account of artistic illusion in dialogue with the psychology of perception, which he deemed essential to an investigation into the modes of production and interpretation of artworks. The image of the duck-rabbit made famous (at least among philosophers of science)[10] by Thomas Kuhn, for instance, features in the introduction of *Art and Illusion* (first published two years earlier than *The Structure of Scientific Revolutions*), to show the impossibility of detangling perception from interpretation, and to highlight the crucial role of learning and expectation in making sense of the ambiguity that characterises what is before one's eyes – in real life just as in art making. 'Painting is an activity', Gombrich claimed later on in the book, 'and the artist will therefore tend to see what he paints rather than to paint what he sees'. (Gombrich [1960] 2002, p. 73)

What perhaps attracted Feyerabend's attention towards *Art and Illusion* was the critique of the legacy of empiricism in art making, which Gombrich pursued with an eye to the works of his lifelong friend Karl Popper.[11] 'The inductivist ideal of pure observation has proved a mirage in science no less than in art', Gombrich pointed out, explicitly invoking Popper: 'Every observation, as Karl Popper has stressed, is a result of a question we ask nature, and every question implies a tentative hypothesis' (Gombrich [1960] 2002, p. 271). This account of the conjectural nature of observation underpins Gombrich's appropriation of Popper's method of conjectures and refutations, and its application to the domain of art:

> We look for something because our hypothesis makes us expect certain results. Let us see if they follow. If not, we must revise our hypothesis and try again to test it against observation as rigorously as we can; we do that by trying to disprove it, and the hypothesis that survives that winnowing process is the one we are entitled to hold, *pro tempore*.
>
> This description of the way science works is eminently applicable to the story of visual discoveries in art. Our formula of schema and correction, in fact, illustrates this very procedure. You must have a starting point, a standard of comparison, in order to begin that process of making and matching and remaking which finally becomes embodied in the finished

---

[10] On the history of the duck-rabbit before Kuhn, from its creator Joseph Jastrow to Ludwig Wittgenstein, see Viola (2012). On the relationship between science and art in the first manuscript of Kuhn's *The Structure of Scientific Revolutions*, see Pinto de Oliveira (2017).

[11] Gombrich had been instrumental – among other things – in the publication of *The Open Society and Its Enemies* (1945). For recent critical appraisals of the relationships between Popper and Gombrich, see Hemingway (2009) and Schneider (2009). For recent reappraisals of Gombrich's place and influence on the field of history of art, see Wood (2009) and Mount (2014).

image. The artist cannot start from scratch but he can criticise his fore-
runners. (Gombrich [1960] 2002, pp. 271–272)

Art 'making' thus takes the form of a visual conjecture or hypothesis
grounded in conventions. Conventions, in turn, serve as standards of
comparison for any pictorial innovation introduced by the artist with
a new representation. The new pictorial schemata are then 'matched'
against the world and corrected until the image satisfactorily resembles
the portion of reality singled out by the artist. Figurative realism, for
Gombrich, is the hard-won result of this process of trial and error – the
successful matching between what artists make and what they *expect* to
encounter in their field of perception, which is itself shaped by inherited
conventions.

Although Gombrich remained somewhat ambiguous on this point, his
account of representation in relation to the psychology of perception is
neither constructivist nor entirely conventionalist.[12] His main point in *Art
and Illusion* is that there is some kind of 'factual' content to our perceptions, but that content is inherently ambiguous – and so our attempts at
rendering it in a pictorial form are inevitably in the form of conjectures,
formulated with the aid of a set of expectations. Ambiguity, in turn, for
Gombrich 'cannot be seen – it can only be inferred by trying different
readings that fit the same configuration' (Gombrich [1960] 2002, p. 264).
This ambiguity characterises the stage of 'making' pictorial conjectures
about the world as much as the stage of 'matching' those conjectures to the
ways in which the world is experienced from a particular perspective. 'The
world does not look like a picture but a picture can look like the world'
Gombrich (1972, p. 138), explained in a later reflection on the key message
of *Art and Illusion*. And yet that resemblance is an achievement of representation, not a relationship dictated by a unified and immutable reality.

In the long footnote to 'Problems of Empiricism', Feyerabend singles
out what is probably Gombrich's best-known example to illustrate the
conjectural nature of making, and the role of conventions as the starting
point for the rendering of unfamiliar objects in art. Albrecht Dürer's 1515

---

[12] And yet he is often lumped in the conventionalist camp, alongside Nelson Goodman. This is
because in *Languages of Art* Goodman himself co-opted Gombrich into supporting his own
conventionalist cause (Goodman 1976, p. 7). But Gombrich was adamant to distance his approach
especially from the kind of nominalism underpinning Goodman's conventionalist approach: 'He
rather misunderstood my book. He interpreted it as completely "conventionalist"', he explained in
conversation with Didier Eribon (Gombrich 1993, p. 112). For further details on Gombrich's
response to Goodman, see Gombrich (1972), which incidentally also contains a discussion of
Brunelleschi's perspectival rendering of the Baptisterium – an example taken up later by
Feyerabend, as I show later.

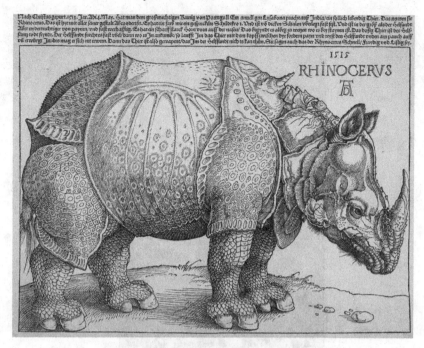

Figure 1.3 Albrecht Dürer, *Rhinoceros* (c. 1515).
© The Trustees of the British Museum. All rights reserved.

iconic rhinoceros woodcut (Figure 1.3) is an instance of conjectural 'making' that crystallised into a stylistic convention in its own right. Drawn from second-hand evidence (despite being presented in the caption as made 'from life', which was a common trope at the time) and supplemented with representative conventions applied to other exotic animals, Dürer's rhinoceros is covered in a thick armour and presents a smaller spurious horn at the top of its neck. Gombrich ([1960] 2002, p. 71) highlights that this image became the standard model for representations of the animal in natural history texts, even in the face of conflicting evidence from observational reports, and even after explorers and natural historians pointed out the difference between the (one-horned) Indian rhinoceros and the (two-horned) African rhinoceros (see Figures 1.4 and 1.5).[13] This is a key point that Feyerabend illustrates with images in his footnote.

---

[13] I have explored the relation between Dürer's iconic representation and the work of the eighteenth-century artist (and rhino-obsessed illustrator) Jan Wandelaar, whose representations are reproduced

Figure 1.4 Jan Wandelaar, 'the rhinoceros as it had been commonly depicted',
in Kolb (1727, p. 189). Allard Pierson, University of Amsterdam, AB 072:03.

Contrasting Dürer's original image with an eighteenth-century painting
and a photograph of an African rhinoceros, Feyerabend adapts Gombrich's

in Figures 1.4 and 1.5, in my Ambrosio (2015). Wandelaar was indeed one of the early illustrators who recognised the difference between African and Indian varieties, but he was also one of the first illustrators who attempted to subvert the established pictorial conventions by liberating the image of the rhino from its traditional armour. His 1727 illustrations of the rhinoceros, contained in a treatise on the flora and fauna of the Cape of Good Hope, would have greatly amused Gombrich and Feyerabend alike. Asked to produce an image in the manner of Dürer, Wandelaar made two different engravings: one labelled 'the rhinoceros as it had been commonly depicted' (Figure 1.4) and the other labelled 'the rhinoceros according to this description' (Figure 1.5). See Ambrosio (2015, p. 123) and Rookmaaker (1976, p. 88). In a Feyerabendian spirit, I had originally placed Jan Wandelaar's images in this footnote. Alas, due to 'possible issues with repagination in certain e-formats' (personal communication), the twenty-first century digital press can no longer cope with images in footnotes, as it did in the case of Feyerabend's 'Problems of Empiricism' in the 1960s. I imagine Feyerabend would be amused but not particularly surprised by this!

Figure 1.5 Jan Wandelaar, 'the rhinoceros according to this description',
in Kolb (1727, p. 190). Allard Pierson, University of Amsterdam, AB 072:03.

views to his main claim against the empiricist variant of naïve representationalism: 'we find that conventions are not used only in the absence of the object, *but exert their influence even when a direct account of the visible object is attempted*' (Feyerabend 1965, p. 221 fn. 8).

Feyerabend's use of Gombrich in 'Problems of Empiricism' remains sketchy and ambiguous, but it seems that *Art and Illusion* provided him with visual and historical ammunition in support of the claim that the stability and uniformity that 'radical empiricists' in his time attributed to the contents of perception was itself at best a conjecture. Indeed, earlier on in the very same footnote to 'Problems of Empiricism' he states: 'what we receive from the outer world (and from the so-called "inner" world) are certain *clues*, which most of the times are pretty vague and indefinite. Perception is the result of the reaction of the organism to these clues' (Feyerabend 1965, p. 220 fn. 8). Gombrich's account of making and matching provided him with a visual and psychological counterpart for the claim, advanced by his pragmatic theory of observation, that perceptions and sensations *become* observational reports when they are interpreted in light of theory. Just like images in Gombrich, observational reports are conjectures about the world, formulated with the aid of conventions and in a particular *style*. But what Gombrich provided him with was also an initial way to rethink the relationship between sense data, observation and representation in dynamic terms, a point that would be crucial to Feyerabend's later critique of 'imitative philosophies' of art and science alike in *Conquest of Abundance*.

While 'Problems of Empiricism' invokes the question of pictorial styles specifically in relation to observation, in the 1980s, Feyerabend's interest in art shifted towards a sustained comparison between science and art on historical grounds. Interestingly, in turning explicitly to the question of styles in their historical contexts, Feyerabend drew on a source in art history that had constituted the very foil against which Gombrich shaped his entire approach: the Austrian art historian Alois Riegl. It is to Riegl's influence on Feyerabend that I turn in the next section.

1.4   Science as Art

In later writings, Feyerabend's use of art extends beyond a critique of empiricism, in line with the turn towards pluralism *in society* that characterised his works in the late 1970s and 1980s (Brown and Kidd 2016, p. 3). Here, art is one of the non-scientific traditions that Feyerabend uses to 'defend society' from the authority of science. These are the years in which Feyerabend describes his philosophy as 'relativist', insofar as it places science as one tradition among others (ibid). It is to this phase that the book *Science as Art* (1984) belongs, its core message being that science is not different from the arts, as the practices and outcomes of both fields are contingent upon the deliberate choice of particular *styles*. The choice of a style for Feyerabend is a social act in the sciences as much as it is in the arts, and the analogy between the two fields aims precisely at fleshing out how criteria of truth, reality, success and verification are internal to the particular style that communities decide to adopt at a certain time in history.

The beginning of *Science as Art* will look familiar to readers acquainted with *Conquest of Abundance*. The book opens with the description of a painting of legendary fame: Brunelleschi's depiction of the Baptisterium in Florence, celebrated as the first systematic application of perspective to painting. Feyerabend was well aware that perspectival rendering was already a feature of Classical art, but the main point in this section of his book is to demonstrate how Brunelleschi contributed to elevate perspective to the status of a science, through the application to painting of precise geometric and optical rules. He achieved this through a simple demonstration, which for Feyerabend exhibits all the characteristics of a scientific experiment (Feyerabend 1984, p. 96). First, Brunelleschi produced an image of the Baptisterium as seen from a precise spot by the doors of the nearby Cathedral of Santa Maria del Fiore. Having drilled a little hole in the centre of the painting, he held it about five feet from the ground, standing exactly in the place from which the image was taken. He then held a mirror just across the painting, so that it would produce a reflection of the depiction of the Baptisterium, which could be seen through the hole at the centre of the painting. At this stage of the experiment, Feyerabend notes, Brunelleschi saw a combination of art and reality: the mirror reflected the painting of the Baptisterium in its lower half, and the sky and clouds above Florence in its upper part (ibid). He finally removed the mirror, and the image before his eyes remained unchanged – except that it was the 'real' Baptisterium that he now saw through the hole in the painting.

As I will show in the next section, Feyerabend will revisit (and partially reinterpret) Brunelleschi's experiment in Chapter 4 of *Conquest of Abundance*. In *Science as Art*, the experiment functions as a springboard to think historically about the possible origins of the 'naïvely imitative view' of art, and how it might have gained momentum and credibility precisely through its relationship with the sciences. It is at this point that Feyerabend introduces a landmark study in the history of artistic styles: Alois Riegl's *Late Roman Art Industry* (*Spätrömische Kunstindustrie*) ([1901] 1985).[14] Riegl's work resonated with Feyerabend's argument precisely because it questioned the common-sense view of linear progress towards increasing fidelity to nature, culminating in the naturalistic triumph of perspective. The core message of *Late Roman Art Industry* was that periods of alleged artistic decadence, traditionally dismissed by art historians, should be studied in their own right and regarded as characterised by their own distinctive and coherent styles.[15] A member of the Vienna

---

[14] Drawing (cautiously) on Feyerabend's autobiography, Ginzburg (1998, pp. 43ff) suggests that he might have been exposed to Rieglian ideas since at least the 1940s.

[15] Riegl's works and their importance in the historiography and methodology of late Roman art and archaeology have seen a revival since the 1990s. The two works in English that contributed to give new visibility to his works are Olin (1992) and Iversen (1993). Interestingly, Olin spotted the German edition of Feyerabend's *Science as Art*, which she briefly mentioned in a footnote as an example of philosophers' and scientists' recent interest in 'Riegl's presumed cultural relativism' (Olin 1992, p.191

School of Art History, and a proponent of formal analysis as a distinctive art historical methodology, Riegl argued specifically against the then widespread view that late Roman and early Christian art, sculpture, architecture and crafts were the mere residues of Classical art, stripped of its pagan connotations and copied by inexperienced artisans. The historiography of art of the time saw this period of artistic decay as mirroring the decline of the Roman Empire; against this view, Riegl argued that the artistic production of the time should be regarded instead as expressing a distinctive and coherent *Kunstwollen*, variously translated as 'artistic volition' or 'will-to-art', informing different artistic periods and styles. It was this 'will-to-art' internal to each style, and not an increased fidelity to nature, which should be the basis against which styles themselves and their peculiar characteristics should be judged.

Riegl's concept of *Kunstwollen* is still the object of divergent interpretations among historians of art. Margaret Iversen interprets it as a response to naïve empiricist as well as determinist, functionalist and materialist tendencies in the history of art. 'Its emphasis on will', she claims, 'was meant to retrieve agency in artistic production from the domain of causal explanation' (Iversen 1993, p. 6). Jaś Elsner (2006, pp. 748ff), along different lines, interprets it as an expression of Riegl's empiricism, and precisely as a methodological solution to the question of how empirical observations of particular objects in the history of art can lead to convincing historical generalisations. Riegl's *Kunstwollen* 'is encapsulated by the struggle between the artist and the limitations imposed by the material he works on and his own technical capacities' (Elsner 2006, p. 750). As such, the concept applies to individual works of art as well as to art in general – showing that special cases and typical examples share a fundamental structure.

However, neither of these interpretations was the received view of Riegl's *Kunstwollen* in Feyerabend's time. Instead, what Feyerabend might have had access to was the scathing critique of Riegl put forward by Gombrich in the introduction of *Art and Illusion*, which is worth quoting directly:

> There is a touch of genius in the single-mindedness by which Riegl tries by one unitary principle to account for all stylistic changes in architecture, sculpture, painting, and patternmaking. But this single-mindedness, which he took to be the hallmark of a scientific approach, made him a prey to those

fn.10). For a sophisticated account of the empiricist roots on the concept of *Kunstwollen*, see Olin (1992) and Elsner (2006).

prescientific habits of mind by which unitary principles proliferate, the habits of mythmakers. The . . . *Kunstwollen* becomes a ghost in the machine, driving the wheels of artistic development according to 'inexorable laws'. (Gombrich [1960] 2002, p. 16)

If this critique has a distinctive anti-Hegelian flavour, it is precisely because Gombrich interpreted Riegl in teleological, historicist terms. 'By inculcating the habit of talking in terms of collectives, of "mankind", "races" or "ages", it [Riegl's kind of history of art] weakens resistance to totalitarian habits of mind' (Gombrich [1960] 2002, pp. 16–17). And if the charge of falling prey to 'prescientific habits of mind' sounds Popperian, it is because Popper was, indeed, the main drive behind Gombrich's vitriolic critique: only a paragraph later Gombrich invokes *The Poverty of Historicism* in support of his assessment of Riegl's methodology.

In *Science as Art*, Feyerabend does not seem particularly worried by the potentially historicist overtones of Riegl's approach (though he would reconsider his use of Riegl later, in *Conquest of Abundance*). Instead, he interprets Riegl's concept of *Kunstwollen* as a claim about the internal coherence towards which each style strives,[16] and from there he derives a series of arguments against linear conceptions of progress to be carried over from art to science. Late Roman architecture, for example, presents a distinctive conception of space: 'It recognised space as a cubic material quantity', Feyerabend (1984, p. 116) explains citing directly Riegl, '[and] it differs in that from ancient Near Eastern and classical architecture; but it does not recognise it as an unlimited shapeless quantity – which makes

---

[16] Feyerabend's adoption of the term resonates with at least two characterisations of styles widely discussed in philosophy of science. One is Ludwik Fleck's ([1935] 1981) concept of *Denkstil* or 'thought style', the collective set of precepts, acquired by habit, which act as constraints upon scientific observation. As Lorraine Daston (2008) has noticed, this goes further than the common appeal to the theory-ladenness of observation in traditional philosophy of science: 'Fleck was concerned with how perception forged stable kinds out of confused sensations' (Daston 2008, p. 100). This characterisation seems to chime with Feyerabend's views about the conjectural nature of perception, which I discussed in the previous section. A second characterisation of styles immediately relevant to Feyerabend's discussion in *Science as Art* is Ian Hacking's work on styles of reasoning (subsequently relabelled 'styles of thinking and doing' (Hacking 1982; 1992; 2012). While Feyerabend never developed a taxonomy of styles (which Hacking himself famously adopted from Alastair C. Crombie's 1994 work, *Styles of Scientific Thinking in the European Tradition*), what he seems to be looking for in the concept of *Kunstwollen* could be regarded as analogous to the description of styles as 'autonomous and self-authenticating', which is distinctive of Hacking's (e.g. 1992, pp. 13ff; 2012, p. 605) characterisation. And indeed, Hacking himself later acknowledged Feyerabend's influence on his styles project, which he described retrospectively as driven by Feyerabendian 'anarcho-rationalism' (Hacking 2012, p. 600). For an illuminating analysis of the parallels between Riegl and Hacking's respective accounts of styles, see Kwa (2012).

it different from modern architecture' (Riegl [1901] 1985, p. 43). Sculpture
obeys similarly consistent stylistic principles: the reliefs of the Arch of
Constantine in Rome, Feyerabend notes, referring to one of Riegl's most
celebrated examples, are the culmination of the 'spatial isolation of the
figures' (Feyerabend 1984, p. 117; Riegl [1901] 1985, p. 53) towards which the
late Roman style strove. The same can be said of late Roman portraiture,
and especially of the rendering of individual figures in late Roman mosaics,
where all the parts are reproduced in equal measure, independently of how
they are positioned in space.

The first lesson Feyerabend draws from Riegl in *Science as Art* is that
differences in styles are expressions of particular artistic intentions, and as
such they defy any characterisation in terms of linear progress or decline.
The triumph of perspective, celebrated by Vasari as the culmination of
figurative realism, is only one among such expressions. The discussion of
the Baptisterium is a case in point. Brunelleschi's experiment, Feyerabend
argues, involves the comparison of *two* artefacts, both expressions of the
perspectival style characteristic of the Renaissance: the painting of the
Baptisterium, and the Baptisterium as it *appeared* to an observer located
in a precise position and trained in the particular style afforded by perspective (Feyerabend 1984, p. 119). This initial consideration for
Feyerabend has implications and potential for a further application of
Riegl's lessons to the case of science. As in the visual arts, so in the sciences
styles allow us to go beyond the notions of linear progress and decline. But,
the payoff of Riegl's account is especially visible, Feyerabend notes, when
we apply his views to some of the major themes that have concerned
philosophers of science for a long time: reality and truth, success and the
verification of scientific theories.

Feyerabend's application of Riegl to science aims to question the naïve
representationalist core of contemporary accounts in philosophy of
science, showing how both empiricists and realists ultimately converge in
postulating *one* conception of reality and *one* mode of representation that
can satisfy it. But any attempt at defending this very basic variety of
monism, Feyerabend argues, fails at taking the debate any further than
Riegl. It could be argued, for example, that pure mathematics is the closest
scientific equivalent of the role of forms that Riegl places at the core of
different styles. Just like forms in Riegl, pure mathematics provides the
scientist with a range of representational techniques that afford an
'advanced investigation of reality' (p. 121). What the realist would point
out here is that scientists can choose among competing representations, but
that process of choice ultimately still aims at isolating the single true

representation that best captures reality. We can concede, Feyerabend continues, that perhaps defenders of the authority of science would want to apply a more nuanced version of Riegl's account of art to the case of the sciences. In this nuanced view, we could capitalise on the multiplicity of representational styles in science to suggest that the aim is to *compare* them, just like we can compare the features of Renaissance and late Roman styles and arrive at a judgement over where one succeeds and others fail. By this account, for instance, late Roman art fails in its representation of space, which is far more successfully and accurately captured with the introduction of perspective. But this more nuanced suggestion, Feyerabend argues, still implies the existence of a neutral point of view from which both styles can be objectively compared, which is precisely what Riegl's notion of *Kunstwollen*, as interpreted by Feyerabend (pp. 122–123),[17] calls into question. Appealing to an external notion of reality as an objective criterion will not do either, as it assumes that artists (or scientists) already possess a predefined notion of reality, towards which they orient their practices.

The same line of argumentation applies to philosophers of science's characterisation of the conditions of verifiability of scientific predictions. Affirming that observations can lend more or less support to – or conclusively falsify – a prediction, is itself an assumption built on the idea that there is a single, stable perceptual world against which our theories can be tested. Drawing on the transition from Aristotelian to Galilean physics, Feyerabend shows that even such a simple assumption is an oversimplification: conditions of verifiability are themselves subject to styles. Thus, for instance, Aristotelian observations aimed to ascertain qualities where Galilean physics takes a quantitative approach; in the former, predictions have a modest role whereas the latter is almost entirely built upon them, and so on. The upshot of Feyerabend's discussion is that no matter how we try to use reality – in this narrow formulation – as the objective grounding for a discussion of styles in art and science, we fall back into Riegl's position, which he explicitly labels as 'relativism' (p. 127): styles can only

---

[17] Jaś Elsner presents an alternative view on this rather problematic aspect of Riegl's *Kunstwollen*. On one level, he shows that Riegl did subscribe to a notion of progress – albeit one that is incommensurable with contemporary criteria. Contrasting the late Roman *Kunstwollen* with that of the Flavians and Trajan, Riegl states that the former 'constitutes progress and nothing other than progress. Judged by the limited criteria of modern criticism, it appears to be a decay that did not exist, but modern art with all its advantages would never have been possible if late Roman art . . . had not led the way' (Riegl [1901] 1985, p. 11). In the same vein, Elsner shows that Riegl himself attempted to compare the modern *Kunstwollen* of the nineteenth century with the late Roman one, thus at least implying that comparison was possible. See Elsner (2006, p. 751).

be chosen, developed and evaluated according to criteria internal to themselves.

The upshot of Feyerabend's application of Riegl to science is that reality, truth, success, verification are all ultimately historically contingent *choices* of particular modes of operating carried out by human beings in precise social contexts. While this was only acknowledged in the case of the arts, the inescapability of Riegl's conception of style when applied to scientific practice is a legitimate ground, Feyerabend argues, to reconsider the *sciences as arts* in this modern sense (p. 156).

1.5   Imitating an Abundant World

It is to Riegl that the late Feyerabend returns, revisiting his account of style in a critical spirit. Where *Science as Art* emphasises the inescapable nature of Riegl's relativism, *Conquest of Abundance* reconfigures Riegl's contribution, highlighting some crucial questions that his account leaves open. 'Riegl's observations are a valuable corrective to crude progressivism', Feyerabend here states, but 'his positive ideas are another matter' (Feyerabend 1999, p. 93). Here Feyerabend returns to the question of imitation, which Riegl would consider as dependent on what artists assume reality to be in different historical contexts. While this seems to reconcile imitation and stylistic pluralism, Feyerabend notes that it does so 'on the basis of an arbitrary and badly founded assumption' (p. 93): the close correlation between artists' assumptions about reality and the practices we identify as 'imitation' is merely postulated by Riegl. In this respect, his ideas fare no better than the naïve realist's assumptions about the postulation of the 'worldview-independence . . . of reality' (p. 93).

Feyerabend scholars have pointed out that *Conquest of Abundance* exemplifies the quest for a metaphysical counterpart to radical pluralism (Kidd 2010, chapter 8; Tambolo 2014) and the broader return to realism which are distinctive of the late Feyerabend (Kidd, 2010; Brown 2016; and Kidd 2016) . In this last section, I want to argue that this shift is also at the basis of his return to a richer notion of representation, and to his renewed interest in a notion of imitation as a *process*. Where *Science as Art* drew on (Feyerabend's own interpretation of) Riegl's notion of self-consistent styles to advocate the necessity and coexistence of a plurality of representations (with scientific theories, models, explanations and predictions obeying their own internally coherent stylistic criteria), in *Conquest of Abundance*, the metaphysical premise of Feyerabend's account of representation changes. This metaphysical premise is sketched in the opening of the book, now a much-quoted passage by Feyerabend:

> The world we inhabit is abundant beyond our wildest imagination. There are trees, dreams, sunrises; there are thunderstorms, shadows, rivers; there are wars, flea bites, love affairs; there are the lives of people, Gods, entire galaxies. (Feyerabend 1999, p. 3)

We can handle only a fraction of this abundance, Feyerabend states, and this is to a certain extent a blessing: 'a superconscious organism would be paralysed' (pp. 3–4). But, this also means that to make any sense of the world, abundance needs to be simplified, either in thought via abstraction or by actively interfering with it via experiments. In both cases, a large part of that abundance is 'blocked off', and what remains is considered amenable to investigation – and is referred to as 'the real' (p. 5).

The metaphysical thesis of an abundant world has been investigated by Feyerabend scholars from various angles. Kidd (2010, chapter 9; 2012) has connected it to the late Feyerabend's characterisation of the ineffability of reality (Feyerabend 1999, pp. 195–196), and related it to the writings of the Neoplatonist Pseudo-Dionysius the Aeropagite. In analogy with Pseudo-Dionysius' characterisation of God's names, explicitly acknowledged by Feyerabend (p. 195), reality as such is ineffable, but 'depending on our approaches it may respond in a variety of comprehensible ways' (p. 196). Tambolo (2014) sees abundance as part of Feyerabend's articulation of a broader argument about the *limited pliability of reality*, a thesis entailing two claims: reality, as Feyerabend states, is *pliable* (p. 145) – it can be moulded by our descriptions in an indefinite number of ways, each of which answers particular questions we are asking nature. But, this pliability is limited by *resistance*: 'some constructions', Feyerabend explains, 'find no point of attack ... and simply collapse' (p. 145). Matt Brown (2016) has brought these accounts together in a systematic reconstruction of Feyerabend's metaphysics, showing that 'the abundant world' is a complex but coherent thesis articulated throughout *Conquest of Abundance*. Brown characterises it as a form of ontological pluralism aimed at refuting 'scientific materialism', a family of realist accounts that subscribe to, or combine in various ways, mind-independence, taxonomical monism (the world has a single, coherent and uniform structure of entities and processes), ontological reductionism (higher order structures can be reduced to the properties and relations of lower order ones) and physicalism (in the formulation that the basic structure of the world is physical) (Brown 2016, p. 143).

All these accounts agree in characterising Feyerabend as denying that there is a single, unique scientific description or representation that will exhaust the 'richness of Being' (which Feyerabend uses as synonymous with reality). Kidd (2012) and Brown (2016) also show that Feyerabend's late metaphysics aims at eliminating the artificial (neo-Kantian) distinction between appearances and reality: the distinction is only a product of the process of simplifying abundance, and it is itself an 'invention' of Western philosophy. Instead, Feyerabend differentiates between Being, which he characterises as ineffable, resisting but pliable (as in Tambolo 2014) and 'manifest worlds', which are the products of how Being responds to our beliefs, goals, interests and practices (Feyerabend 1999, p. 204). As Brown (2016, p. 147) explains, manifest worlds are neither ideal entities nor phenomenal worlds; they are just evidence of our interactions with Being (or with an abundant reality). 'Inhabitants of a particular manifest world', Feyerabend explains, 'often identify it with Being. They thereby turn local problems into cosmic disasters' (Feyerabend 2016, p. 204). The evidence we have of our *interactions* with an abundant reality should not be confused with Being *itself*: this confusion results in mistakenly considering these fragmentary worlds as complete representations of an objective, mind-independent reality (Kidd 2012, p. 369). On the contrary, Feyerabend argues, 'the manifest worlds themselves demonstrate their fragmentary character; they harbour events which should not be there, and which are classified away with some embarrassment (example: *the separation of the arts and the sciences*)' (Feyerabend 1999, p. 204, emphasis added). Feyerabend's cursory comment is telling: the separation of the sciences and the arts is yet another result of the fragmentary character of a particular manifest world, in which they are regarded as separate domains. To a certain extent, his work on the relationship between these two fields is thus itself an attempt at recovering a mode of engagement with Being that had been suppressed in the process of simplifying abundance.

It is this metaphysics of abundance that underpins, I claim, Feyerabend's revised account of the role of imitation in representation. Reality, or Being, is abundant. It is an open domain of possibilities, which are not exhausted by our descriptions or representations. At the same time, descriptions and representations are among the practices and processes we carry out to come into contact with Being, and as such they are part of particular manifest worlds that evidence our interactions with Being. They are inevitably fragmentary, partial and perspectival – a point that Feyerabend develops in detail in chapter 4 of *Conquest of Abundance*.

Immediately after revisiting his position on Riegl, Feyerabend moves on to reconsider Brunelleschi's visual experiment with the Baptisterium. Just like in *Science as Art*, he stresses that Brunelleschi's experiment shares many of the features of scientific experiments: the perspectival construction of the painting obeys rigorous rules derived from architectural practice; the rendering of the Baptisterium is performed under very specific conditions. Again, in line with *Science as Art*, Feyerabend states that here Brunelleschi is comparing two objects: the painting of the Baptisterium and 'something else. This "something else" was not a building; it was an *aspect* of a building ... Brunelleschi chose an aspect that suited his purpose' (Feyerabend 1999, p. 100). But at this point, Feyerabend introduces a theoretical angle which was not present in his earlier account of the experiment, and which is directly related to the renewed interest in imitation *as a process* that emerges in *Conquest of Abundance*. In a sense, Feyerabend argues, Brunelleschi's is an attempt at representing by imitating reality. But to understand *how* it is necessary to switch from a narrowly *visual* to a broader, *theatrical* sense of imitation:

> If we want to say that Brunelleschi imitated reality then we have to add that this reality was manufactured, not given ... The best way to describe the situation is by saying that Brunelleschi built *an enormous stage*, containing a pre-existing structure (the Baptisterium), a man-made object (the painting), and special arrangements for viewing or projecting both. The reality he tried to represent was produced by the stage set, the process of representation was part of the stage action, it did not reach beyond it. (pp. 100–1)

In his reconstruction of Feyerabend's account of Brunelleschi's experiment, Matt Brown (2009) compares Feyerabend's staging metaphor to the kind of scientific perspectivalism advocated by Ronald Giere (2006). Just like in Giere's view, Brown argues, 'man-made objects (paintings, theories) are compared with the World only through projections ... Theoretical principles must be transformed into representational models, and the scientist must generate models of data in order to make a comparison' (Brown 2009, p. 217; cf. also Giere 2006, chapter 4). But Brown takes the Feyerabendian account of representation one step further, showing that the staging metaphor entails also the creation of an *audience*, 'the mostly unspecified agent in Giere's account' (Brown 2009, p. 217). In Feyerabend's reconstruction, Brown shows that theories are not compared with the world. Instead, what is compared are two functional artefacts: representational models and models of data. Moreover, Brown continues, 'the similarity or fit between these two objects, is not an abstract relation, but it is an act carried out by agents fulfilling another functional role in the process of representation, the audience' (p. 217).

As in every theatrical metaphor (as well as in actual theatrical performances), 'audience' here can be constructed in two ways. One is passive: the staged performance 'directs' the audience's attention in particular ways. But this is clearly not the sense intended by Brown, who states that scientists qua audiences have a functional role in the process of representation. The other is active: the staged perspective, as in a play, dynamically *moves* the audience towards new possibilities (the audience can make choices and take decisions on how to interpret and act upon the staged representation). It is this second, active sense of 'audience' that Brown seems to be hinting at, and it is this dynamic sense of 'staging' as involving audiences and their agency, I claim, that can be used to make sense of the Feyerabendian way of reconsidering the role of imitation as a *process* in *Conquest of Abundance*.

The dynamic sense of imitation I want to attribute to the late Feyerabend is one of Aristotelian legacy, and it is precisely the notion of *mimesis* that Aristotle saw at work in drama.[18] Indeed, the Aristotelian version of the imitative view is cursorily acknowledged by Feyerabend in Chapter 4 of *Conquest of Abundance*: 'The imitative view ... was developed by Aristotle (tragedy imitates deep-seated social structures and is therefore "more philosophical" than the most painstaking historical account)' (Feyerabend 1999, p. 92; cf. also 1987, p. 129). In *Farewell to Reason*, Feyerabend proves to be well acquainted with the subtleties that characterised the concept of *mimesis* in antiquity and through the centuries. In the space of two pages, he presents the trajectory of *mimesis* from Book 10 of Plato's *Republic*, where artists are criticised 'for imitating the wrong entities, ... for making deception ... part of their imitative techniques, and for arousing emotions' (p. 128), to Aristotle's revival of the concept in his theory of drama 'contained in his magnificent *Poetic*' (p. 128), to the revival of *mimesis* in the Renaissance and all the way to the invention of photography in the mid-nineteenth century (p. 129). What this history shows, Feyerabend claims, is that 'imitation is a complex process that involves theoretical and practical knowledge (of materials and traditions), can be modified by invention, and always involves a series of choices on part of the imitator' (p. 130). The staging metaphor in chapter 4 of *Conquest of Abundance* capitalises on the subtle differences that characterised the history of *mimesis*, and of which Feyerabend was clearly aware.

---

[18] See in particular *Poetics*, Book 9.

Imitation is one of the many practices included in our manifest worlds, one of the ways in which we come into contact with a pliable and abundant reality. As such, there is not a *single*, correct way of imitating reality. There is no single, unified concept of mimesis. And there is no linear progress towards increasingly more faithful representations, because the criteria according to which we imitate are *contingent* upon our interactions with reality and our interpretations of how reality responds to our interests, beliefs and practices.

In his monumental *The Aesthetics of Mimesis* (2002), Stephen Halliwell proposes that the history of this philosophically contested concept is characterised by the tension between two conceptions: on one hand, there is a view of *mimesis* as 'outward-looking', or committed to illuminating a world that is (partly) accessible outside art, and by whose norms art can be tested and judged. On the other, there is a notion of *mimesis* as 'inward-looking/world simulating/world creating'. In this second account of Aristotelian legacy, *mimesis* is construed as the creator of an independent 'artistic heterocosm' – a world of its own, but also one which still contains *some* truth about reality as a whole (Halliwell 2002, p. 5). The philosophical history of mimesis, Halliwell claims, is ultimately the history of the perpetual tension between these two views. More importantly, Halliwell shows through painstakingly detailed work on historical texts that the history of *mimesis* shows 'no central commitment ... to the truth-bearing, as opposed to the sense-making, status of mimetic works' (p. 380). This is a claim that Feyerabend would have been sympathetic to – and to a certain extent, it is in this direction that his own overview of imitation 'as a complex process' heads. But, even more, Feyerabend would have been sympathetic to Halliwell's reconstruction of Aristotelian *mimesis*, which, I claim, lurks in the background of his metaphor of representing as 'staging', and is compatible with the metaphysical outlook he adopts in *Conquest of Abundance* more broadly.

In the *Poetics*, Aristotle characterises poetry distinctively as a mimetic art concerned with 'things which *could* be the case, and which are *possible* in terms of probability or necessity' (*Poetics* 9.1451 a37-38, emphasis added).[19] It is this character of possibility that opens up an account of *mimesis* to an interpretation that Feyerabend would find congenial. *Mimesis* in Aristotle is bound and related to intentionality (of the makers, performers and

---

[19] Here I am using Halliwell's own translation of Aristotle's passage, in Halliwell (2002, p. 154). In an earlier translation (Aristotle and Halliwell, 1987, p. 40), Halliwell presents the passage as follows: 'the kind of events which *could* occur, and are possible by the standards of probability or necessity' (emphasis in the original).

audiences of a representation), and as such it cannot be exhausted by sensory correspondence. 'Mimetic likeness', Halliwell states, 'entails an intentionality that is ultimately natural in origin but becomes embodied in culturally evolved and institutionalised forms. This is one reason why not all likenesses are mimetic: not all likeness has the intentional grounding that is a necessary condition of artistic mimesis' Halliwell (2002, p. 156). This account of mimesis, Halliwell continues, is *relational and transformative*: 'Constituted partly by the experience that it opens up for, and introduces in its audience' (p. 161). This Aristotelian notion of *mimesis* brings together creatively makers, performers and audiences of a representation, and is connected by Halliwell to broader perspectival accounts in epistemology. Drawing a parallel with Putnam's (1992) *Realism with a Human Face*, Halliwell argues that this Aristotelian account of *mimesis* is 'a locus of possibilities within a fully human perspective, a perspective that interprets 'reality' through culturally structured but disputable (and amendable) frameworks of beliefs, standards and conventions rather than by *a set of metaphysically absolute reference points*' (Halliwell 2002, p. 376, emphasis added).

Translated in Feyerabend's terms, this would amount to saying that likeness or imitation is not a single, unified category that cuts across all manifest worlds. And indeed, this is one of the core messages of his investigation of Brunelleschi's experiment, which incidentally also took Feyerabend all the way back to the issue of styles: 'interpreting artworks as stage sets' he claims in Chapter 4 of *Conquest of Abundance*, 'provides a precise and useful framework for discussing a variety of assumptions about the scope, function and development of artistic styles' (Feyerabend 1999, p. 101). First, the stage metaphor gives the makers of representation a new sense of agency. With a nod to Gombrich's criticism of Riegl, Feyerabend points out that artists have substantive control over changes in styles: 'Brunelleschi was not swept along by overwhelming historical forces; he prepared every step of his performance' (ibid). In distancing himself from the historicist readings of Riegl, Feyerabend here suggests, between the lines, that it was the local account of agency built into the concept of *Kunstwollen* (rather than its historicist overtones) that attracted his attention in the 1980s. For the late Feyerabend, this is best exemplified in the notion of manifest worlds: rather than driven by an unspecified 'will-to-art', representative styles are themselves the result of our interactions with a pliable reality. Second, this time against a purely conventionalist reading of Gombrich, the staging metaphor dispels the idea 'that it is *mind* and mind alone that imposes a style and that styles are therefore conventions, free from the impediments of the material world' (ibid). The *materiality* of Brunelleschi's experiment mattered: the correct application and use of his experimental apparatus allowed a particular perspectival representation – an aspect of the Baptisterium – to emerge as a result of the interplay between pliability and resistance distinctive of the manifest world in which Brunelleschi operated. But it is also the case, Feyerabend continues, that the representation did not obey *exclusively* physical laws. True, the final picture of the Baptisterium was constructed with the aid of rules that are based, for instance, on approximations of the laws of propagation of light. But, these laws alone do not guarantee that a human, correctly positioned, 'will see things accordingly' (p. 102). Just like in Halliwell's account of *mimesis*, manifest worlds give us perspectival representations 'within a fully human perspective': they express particular configurations of our interactions with a pliable but resistant reality, according to conventions that render those perspectives amendable and amenable to public scrutiny.

This takes us full circle to where we started: to the 'ugly Madonna of Siena', who 'might have caught an element of reality', but as Feyerabend argues, 'this must be established by research, not by metaphysical speculations about 'the nature of reality' (pp. 93–94). And 'research' is precisely what compels us to inscribe representations within the stages – the manifest worlds – in which they are produced, performed and negotiated. As a 'locus of possibilities', imitation is a varied and creative process, which captures the ways in which we interact with reality and reconstitute it on the basis of our interpretations of reality's responses. This is precisely the conclusion to which Feyerabend arrives through his long journey from scientific to artistic representation, and back again:

> Being in the world we not only imitate and constitute events, we also *reconstitute them while imitating* them, and thus change what are supposed to be stable objects of our attention. (p. 128, emphasis added)

1.6 Conclusions

In this chapter, I explored Feyerabend's writings on art and placed them in dialogue with some of the milestones in his philosophy of science. I focussed on his treatment of representation across art and science, tracing the journey that took him from 'Problems of Empiricism' to *Conquest of Abundance*. I showed how his critique of 'naïvely imitative philosophies' of the empiricist and realist variety compelled him to draw on sources in the history of art to formulate a sophisticated, dynamic account of imitation in art as well as science.

I highlighted two interconnected lines of inquiry in Feyerabend's approach to representation. One is the pervasiveness of the issue of artistic styles, which Feyerabend exploited as a springboard to question 'naïvely imitative' views in science. Two influential sources in these early and middle phases of his philosophy were Ernst Gombrich and Alois Riegl. Gombrich's account of making and matching provided Feyerabend with an empirical and theoretical grounding in the methodology of art history, which also offered him a compelling analogy to rethink the relationship between sense data, observation and representation in dynamic terms. Riegl's concept of *Kunstwollen* helped him shift from the individual psychology of perception distinctive of Gombrich's account to a collective, historicised (albeit not 'historicist', in a Hegelian sense) notion of style. I showed that while the early Feyerabend suspended his judgement about the potentially historicist over-tones of Riegl's account, the later Feyerabend showed at least some awareness of the implications of Gombrich's criticism of Riegl's historicism. But, by that point, what Feyerabend was initially looking for in the concept of *Kunstwollen* had been replaced by the more nuanced notion of 'manifest worlds', as expressions and evidence of scientists' (and artists') interactions with a pliable reality.

The second line of inquiry I identified in Feyerabend's journey from science to art is the question of imitation as a conceptual category in its own right and its relation to representation. Here, Feyerabend's ideas aligned once again with the various phases of his philosophy. *Science as Art* drew on the analogy between styles in art and science to expose the flaws of linear notions of scientific progress, implicit both in realist and verificationist accounts of science. In this context, 'naïvely imitative philosophies' formed the core of Feyerabend's attack against a narrow conception of progress construed as increasing fidelity to nature. This attack was in line with his views about the necessity of pluralism in society, and his defence of society from the authority of science. *Conquest of Abundance,* on the other hand, rescued a role for imitation as one possible mode of representation. But, it did so based on the different metaphysical premise of an 'abundant world'. The late Feyerabend found a new space for imitation (though he never reduced representing to imitating) as a dynamic, performative category, and as a mode of reconfiguring a pliable reality while imitating it.

There is much more to Feyerabend's views on art than the account of his changing views on imitation I outlined in this chapter. While his passion especially for Renaissance painting is at least partly known, his writings on theatre, especially on Bertolt Brecht and Eugène Ionesco, have only been very partially explored (cf. Couvalis 1987). The now established scholarship on Brecht's critical reception of Aristotle's *Poetics* (see for example Halliwell 1998, pp. 316ff; Curran 2001; Halliwell 2002, pp. 372–374), for instance, could serve as an interpretative framework to re-read the metaphor of staging in *Conquest of Abundance* and place it in dialogue with Feyerabend's broader and continued interest in the aesthetics of performance. Along complementary lines, Feyerabend's rich writings on art and science are beginning to serve as a conceptual framework for artistic research, especially when engaged in a critical dialogue with science (see for example Magee 2018). Here I tried to open an avenue of inquiry, which could be taken in a multitude of novel directions, especially by scholars who, in a Feyerabendian spirit, will disagree with the interpretation I presented of his account of representation. There is much more to Feyerabend's writings on art: just like his characterisation of reality, they are 'abundant beyond our wildest imagination'.[20]

[20] Acknowledgements: This chapter owes a great deal to the community of Feyerabend scholars who encouraged me to firm down in writing my thoughts about Feyerabend on art and science. I am especially grateful to Jamie Shaw and Karim Bschir for inviting me to contribute to this volume, and for their constructive comments on earlier drafts of this chapter. Almost twenty years ago now, Hasok Chang taught me Feyerabend in the most inspiring and most effective way: by asking me to teach him. I have been a practising Feyerabendian ever since, and never stopped being grateful for it. Ian Kidd kept insisting that I write about Feyerabend on art and science, and proactively made it happen by inviting me as a keynote speaker to the wonderful conference 'Feyerabend 2015: Forty Years "Against Method"', held at Durham University. The participants of the Egenis Research Exchange reading group at the University of Exeter provided invaluable comments to this chapter. I am especially grateful to Gemma Anderson, Benjamin Smart and Sabina Leonelli, for organising the reading group and pre-circulating a draft of this chapter amongst its participants. Niall Le Mage read and commented on every iteration of this chapter, and provided precious technical support with the images. More importantly, he caught me out every time I did not practise the pluralism I preached, proving to be the real Feyerabendian in our home! Last but not least, this chapter is dedicated to my three students and fellow anarchists Adam Holland, Rachel Fong and Paul Magee, whose insightful works and enthusiasm in Feyerabend's ideas made me a better scholar and a better person.

CHAPTER 2

The Coherence of Feyerabend's Pluralist Realism

Hasok Chang

2.1 Overview

Contrary to common impressions, Paul Feyerabend was a defender of realism throughout his career. It is the task of this chapter to give plausibility to that contentious statement. I hope that this exercise will not only contribute to a better and clearer understanding of Feyerabend's philosophy, but also generate helpful insights for those philosophers of science, epistemologists, metaphysicians and practising scientists who continue to wrestle with the question of realism in and about science.

It should not be a surprise to any careful reader of Feyerabend's texts that he advocated a kind of realism strongly in his early work. There are explicit statements to that effect in Feyerabend's own writings, and John Preston devoted a whole chapter to this topic in his well-known exposition of Feyerabend's philosophy (Preston 1997a, chapter 4). But in his later work, Feyerabend's stance would seem to be more in line with the popular image of him as an iconoclastic postmodernist who rejected the arrogance and absolutism often associated with scientific realism. Preston (1997b) states that Feyerabend retreated from realism in his later work. Although there is some truth to this reading, I will argue (see Section 4) that there is a common thread running in Feyerabend's thought concerning realism, though 'realism' may be an inconvenient label because of the ambiguities in the meaning of the term.

It is commonplace to observe that Feyerabend freely changed his mind and even contradicted himself, and it may be unwise for the interpreter to impose too much tidiness on his thought. However, I do think that there was a very coherent direction of development in Feyerabend's thought when it comes to the issue of realism. He began with a rather narrowly defined position on realism, and then went on to broaden and strengthen it as the years went by. I agree with Luca Tambolo's view that even

Feyerabend's last writings were 'fully compatible with a robustly realist view of science', and that the later Feyerabend can be seen as 'a potential ally of sophisticated versions of scientific realism' (Tambolo 2014, p. 197). The key is to understand that his realism was pluralistic through and through. And to understand the development of his views, it is crucial that we lose the common prejudice that pluralism cannot possibly be realist; dispelling that prejudice may be the most useful thing I can contribute to the discussion of Feyerabend's realism. There was no abrupt change in Feyerabend's pluralist realism; it is only that his pluralism became stronger, more explicit and more cogent.[1] In my reading, it is only what Feyerabend would have considered a *perversion* of realism that he explicitly objected to in his later work. The rejection of such 'realism' actually ran through his work from the start, though it is clearer in the later works.

2.2 Feyerabend's Early Advocacy of the Realist Pursuit of Theories

2.2.1 *Feyerabend's Early Stance on Realism*

I begin by acknowledging again my debt to John Preston's work. Not only did he give us the first comprehensive book-length treatment of Feyerabend's philosophy,[2] but he has also provided a detailed and illuminating discussion of Feyerabend's views on realism in that book and in later articles. And in fact I agree with almost everything Preston says about Feyerabend's take on realism, if taken on Preston's terms. But I have a sense that, in the end, he finds Feyerabend's views disagreeable and that this prevents him from getting the maximum possible sense out of Feyerabend's statements. What I can offer here is a different perspective, arising from a more sympathetic engagement.

The first thing to confirm briefly is that the early Feyerabend was an advocate of realism, and explicitly so. Richard Burian (1971, p. 49) goes as far as to say: 'We may label the central doctrine of Feyerabend's positive programme "scientific realism"'.[3] But what exactly did Feyerabend mean

---

[1] In this view about Feyerabend's attitude towards pluralism, I concur with Jamie Shaw (2017), though other experts seem to disagree.

[2] One should also note the earlier publication by George Couvalis (1989), and Richard Burian's (1971) very early PhD dissertation on the work of Feyerabend before *Against Method*.

[3] This is despite the fact that Burian (1971, p. 239) then goes on to give a harsh verdict that Feyerabend's programme was seriously 'deficient' as realism, despite 'his occasional appropriation of the label "realism" for his philosophy.'

by 'realism'? Burian (1971, pp. 49–55) notes that the central point is to take scientific theories seriously, as accounts of the world 'with ontological relevance', which can often reveal aspects of reality that are contrary to the testimony of our senses. What Feyerabend was talking about in his early works fits nicely into the later and now-classic articulation of scientific realism given by Bas van Fraassen (1980, p. 8): 'Science aims to give us, in its theories, a literally true story of what the world is like; and acceptance of a scientific theory involves the belief that it is true. This is the correct statement of scientific realism'. The two main points here are a literal interpretation of scientific theories when it comes to unobservables, and the policy of pursuing the truth about the literally interpreted statements about unobservables.

Feyerabend's early papers (starting with 1958/1981) were quite clear that scientific theories should be given literal interpretations, if possible; in a move against positivism, he advocated the 'realistic interpretation' of theories, according to which 'a scientific theory aims at a description of states of affairs, or properties of physical systems, which transcends experience' (1960/1981, p. 42). He argues against global instrumentalism (which he considered a variety of positivism),[4] and he is also at pains to push back on 'local instrumentalist' arguments, which argue that in some important cases (e.g. quantum theory) it is impossible to give a literal interpretation of the theory. When that appears to be the case, scientists should do something to make a literal interpretation possible. This is what Galileo did with heliocentrism, by undermining the observational facts that seemed to preclude a literal interpretation of heliocentric theory, and by creating a new kind of dynamics in which the motion of the earth could make physical sense.

This is an advocacy of pursuit (much more than it is an assessment of success), as in van Fraassen's definition of scientific realism. Galileo did not know in advance that the heliocentric theory would be able to thrive thanks to the changes he introduced, but Feyerabend judges that what Galileo did was worth trying. This judgement is not tied to our retrospective knowledge that what Galileo did was in the end successful, though that knowledge of success undoubtedly adds rhetorical force to the case. But do all theories deserve such realist pursuit? At least interesting ones do, as Feyerabend hints in *Against Method* (1975a, p 98): 'Galileo is to be applauded here because he preferred protecting an interesting hypothesis to protecting a dull one'. And perhaps that comes to saying that any theory that anyone really cares to pursue is worth pursuing.

---

[4] See Preston (1997a, pp. 70–73) for a critical discussion of Feyerabend's argument here.

2.2.2 Arguments for Realism

Having noted Feyerabend's early advocacy of realism, the next question is what arguments he actually gave for it. Why is it the right thing to do, in his view, to defy even accepted facts and methods to pursue certain theories in a realist way? The most general form of the argument is outlined by Feyerabend as follows, in the 1981 introduction to the first two volumes of his *Philosophical Papers*:

(i) criticism → proliferation → realism

As he notes in this retrospect that in the early papers reprinted in Volume 1, 'this chain is applied to a rather narrow and technical problem, viz. the interpretation of *scientific theories*' (1981a, p. ix). He admits that the three terms occurring in that schema are not precisely defined, 'nor does the arrow in (i) express a well-defined connection such as logical implication'. Rather, it indicates 'starting with the left hand side and adding' various other elements, 'a dialectical debate will eventually at the right hand side' (pp. ix–x).

Feyerabend's various arguments for proliferation are well known, which include the idea that proliferation is good because it facilitates a Popperian sort of criticism of theories. But the next step in schema (i) may seem more mysterious: how is proliferation meant to lead to realism? Preston (1997a, p. 63) gives a helpful summary of Feyerabend's line of thought: 'scientific progress ... is best furthered by theoretical proliferation; and scientific realism leads to a proliferation of theories, while positivism does not. This is clearly a development of Popper's suggestion that positivism is heuristically infertile, that it would produce bad science'. But why is it that realism favours proliferation? I think Preston (1997a, p. 72) is correct to note that anti-realists can also practise proliferation, and in fact anti-realism may even be more conducive than realism to proliferation. But there he is being somewhat uncharitable, in downplaying one important side of Feyerabend's thinking, which comes out famously later in *Against Method*: following the instrumentalist respect for accepted observations would have prevented the Copernicans from maintaining the pursuit of their theories; for such proliferation, a realist attitude for one's theory is required, to the extent of emboldening one to dismantle observations in favour of a realist acceptance of the theory.

Having seen Feyerabend's arguments for realism, one may ask why he cared so much about it. To adapt his famous question about science: what's so great about realism? Without being able to go into this question deeply, I want to leave a few suggestions. Schema (i) does indicate that Feyerabend took the importance of criticism as his starting point. As Jamie Shaw argues, even from his very early work, Feyerabend was concerned to argue against 'rationalism', or 'conception of scientific methodology that provides normative exclusive rules for theory pursuit' (Shaw 2020, p. 2). Anti-rationalism in that sense was an expression of a deeper and general anti-authoritarian and anti-absolutist sensibility that would find fuller expressions in Feyerabend's later work, which I will discuss in detail in Section 5 later.

2.3 But What Is Realism, Exactly?

I commented earlier that the early Feyerabend does not seem to have been too concerned about laying down a precise definition of realism. I could be wrong about that, but at any rate this certainly changes later as he enters into an elaborate consideration of the meanings of 'realism' in his paper on 'Scientific Realism and Philosophical Realism' (1981b), written as an introductory chapter to the first two volumes of his *Philosophical Papers*. This is the 'middle Feyerabend' if you will, who took a retrospective on his earlier publications after unleashing *Against Method* and *Science in a Free Society* on the philosophical world. Although Feyerabend's retrospective views on his own earlier work must be treated with caution, I think it is not insignificant that in the 1981 retrospect (1981b, pp. 15–16) he singled out the two realism papers (1958/1981 and 1964/1981) as 'misleading' and 'dogmatic'.

Feyerabend's early position on realism was riddled with significant tensions. In other people's terminology, the early Feyerabend might have been a 'conjectural realist', or a 'quasi-realist.'[5] Either way, there is discomfort. Conjectural realism would argue for the legitimacy of pursuing various theories because we don't know which one will turn out to be correct, in which case pluralism is only temporary, at least provisional even if forever. Quasi-realism would advocate acting as if our theories were true (even though they are not, or we cannot know if they are, or it makes no sense to say that they are) – in this way, pluralism is only bought at the cost of rendering realism really instrumentalist.

With these tensions in the background, it is interesting to review the different senses of realism that he tried tease apart in the 1981 publication

---

[5] The 'conjectural realism' designation can be found in Preston (1997b, p. S421 and S429).

mentioned earlier. He begins with a statement that is frustratingly inde-
terminate but thereby productive: 'Scientific realism is a *general* theory of
(scientific) knowledge. In one of its forms it assumes that the world is
independent of our knowledge-gathering activities and that science is the
best way to explore it' (Feyerabend 1981b, p. 3). Does this description say
anything in particular, if we accept his view that there really is no well-
defined thing called 'science'? We can only assume that he means by
'science' here (as in other places) the rather illusory thing that exists clearly
in the popular and philosophical imagination, the thing in the name of
which the hegemony of 'Western civilization' is justified. More importantly for our current purposes, note that he says that the familiar realist
idea (the mind-independence of the world, and of the scientific knowledge
of the world) is upheld only in 'one of its [realism's] forms', clearly
implying that there are other forms of realism in which this idea is not
affirmed. And why is he emphasising 'general' in the first sentence quoted?
That has to be a rather clumsily phrased warning against taking aspects of
one particular form of realism as pertaining to all varieties of realism.

He proceeds to distinguish at least three, perhaps four, specific types of
realism. The first was practised emblematically by the Copernicans, for
whom 'the issue is about the *truth of theories*'. Feyerabend adds an interesting observation: 'Claims to truth can be raised only with regard to
particular theories. The first version of scientific realism therefore does
not lead to a realistic interpretation for all theories, but only for those
which have been chosen as a basis for research'. (Feyerabend 1981b, p. 5)
Most scientific realists would want to add some strict rules here about what
kinds of theories should be chosen as a basis for research, while
Feyerabend's own position is much more liberal.

The next type of realism Feyerabend identifies is somewhat puzzling:
'A second version of scientific realism assumes that *scientific theories introduce new entities with new properties and new causal effects*' (p. 6). Did he
mean to suggest that *all* theories do this? Most likely not, as he adds: 'a
direct application of the second version of scientific realism ('theories
always introduce new entities') and a corresponding abstract criticism of
'positivistic' tendencies are too crude to fit scientific practice. What one
needs are not philosophical slogans but a more detailed examination of
historical phenomena'. (p. 7) I am all for engaging with details of history,
but it is not clear what kind of doctrine this second type of realism is. What
seems clear is that this realism is an attitude that scientists choose to take
(or not), concerning particular theories. But, what is not clear is when
Feyerabend thinks how the attitude taken can be justified, in each

particular case. The best interpretation I can give is that he thinks the justification can only consist in the fruits of the effort: for example, the vector potential in electrodynamics was taken in a non-realist way by most investigators; Faraday went against this trend, and he was vindicated in various ways, even by the Aharanov–Bohm effect of the mid-twentieth century. Also note that Faraday's original theory of the 'electrotonic state' does not fit with modern theories, so the situation is complicated: 'A theoretical entity may represent a real entity – but not in the theory in which it was first proposed' (p. 6). Then we have a position somewhat like Ian Hacking's 'entity realism', in which the truth of a theory and the existence of an entity can diverge from each other, and we need to have a separate notion of what makes an entity real; this latter is something that Feyerabend articulates later, as I will discuss in the next section.

The third type of realism, which Feyerabend identifies in the works of Maxwell, Helmholtz, Hertz, Boltzmann and Einstein, is even harder to pin down. Feyerabend calls this the *positivistic version of scientific realism* with conscious paradox: 'making judgments of reality here amounts to asserting that a particular "phantom picture" . . . is preferable to another phantom picture' (p. 10). Phantom picture, or 'inner phantom picture', *Scheinbild* in German, is a term from Hertz to characterise our theoretical pictures or models of reality. But Feyerabend explains that for Hertz the correspondence was not directly between the phantom picture and reality, but between inferences, as follows: 'the logically necessary [*denknotwendigen*] consequences of the picture are always pictures of the physically necessary [*naturnotwendigen*] consequences of the objects pictured'). (Hertz quoted in Feyerabend 1981b, p. 8). Hertz thought that we could judge such correspondence – but how? And even if that were straightforward, how should we deal with the problem of underdetermination? Feyerabend noted in the discussion of the second type of realism that 'theories can be formulated in different ways, using different theoretical entities and it is not at all clear which entities are supposed to be the "real" ones' (p. 6). And, indeed, a key part of Hertz's work on mechanics was to highlight the Newtonian, Lagrangian and Hamiltonian formulations of classical mechanics, and to add another one of his own (Hertz 1894/1899, Introduction). Boltzmann thought that the choice came down to the simplicity of the picture, and that 'Hertz made it quite clear to physicists (though philosophers most likely anticipated him long ago) that a theory cannot be an objective thing that really agrees with nature' (Boltzmann quoted In Feyerabend 1981b, p. 9). But, if we go with such a view, the paradox ceases to be benign: such 'realism' is, on the face of it, indistinguishable from anti-realism. So here, it might appear that Feyerabend's embrace of pluralism takes him away from any kind of realism worth its name; why he would have thought of this as realism becomes clear only when we consider his view of reality articulated in his later works.

After distinguishing these three types of realism, Feyerabend seems to add at least one more: 'The ideas of Maxwell and Mach differ from all the versions I have explained so far. They are also more subtle' (p. 11). So subtle, that Feyerabend did not give us simple characterisations. For Maxwell, the key is the method of analogies, which are good because 'they have heuristic potential' unlike mere mathematical formulae, but 'they don't blind us' like simple-minded physical hypotheses do. That is to say, 'Maxwell wants a conception that *guides* the researcher without *forcing* him into a definite path; that makes suggestions without eliminating the means of controlling them.' (p. 12). But how does this constitute realism? That, I suggest, we can only understand when we see Feyerabend's full-blown version of pluralism discussed in the next section. As for Mach, Feyerabend's reading is highly unconventional, though I think correct. Feyerabend's Mach is not the positivist of popular and philosophical imagination, but an anarchistic scientist pursuing knowledge in any way possible: according to Mach, '*science explores all aspects of knowledge, 'phenomena' as well as theories, 'foundations' as well as standards; it is an autonomous enterprise not dependent on principles taken from other fields.* This idea according to which all concepts are theoretical concepts, at least in principle, is definitely in conflict with the positivistic version of scientific realism'. For Mach, 'even sensation talk involves a "one-sided theory"', and the distinction between theoretical and observational concepts is there, but 'it is regarded as temporary and as being subjected to further research' (p. 13). Boltzmann, too, agrees: 'In my opinion ... we cannot utter a single statement that would be a pure fact of experience' (p. 13). But we seem to have lost the thread by now: how does all *this* count as realism, and how does it go with Maxwell's analogy-based realism? That, again, is also only understandable in the light of Feyerabend's later articulations.

In summary: Feyerabend's 1981 attempt at articulating different meanings of 'realism' raised more questions than it answered. Not only did it reveal again some of the tensions in the pluralist realism stated in his earlier works, but it also created puzzles as to how all the different versions of 'realism' he outlined could be considered realist. I want to argue that these tensions were resolved in the later works of Feyerabend. He achieved this

by making his realism and pluralism both stronger, rather than watering
them down.

2.4   The Later Feyerabend: Pluralistic Reality versus 'Scientific Realism'

2.4.1   Did Feyerabend Retreat from Realism?

In this part of the chapter, my main task is to puzzle out Feyerabend's
stance concerning realism in his later works. As indicated in the introductory section of this chapter, the picture I want to present is one of continual
articulation and development right up to his death in 1994, rather than one
of distinct phases.

As mentioned earlier, Preston (1997b; 2000) argues that Feyerabend in
his later and more notorious postmodernist phase retreated from his
earlier realism. Eric Oberheim (2006, chapter 6) counters this view
vehemently: 'Feyerabend did not retreat from scientific realism. He was
never a scientific realist' (p. 204; see also p. 191). Even though I agree with
the points that he makes about Feyerabend's various statements, I also
think that Oberheim's conception of what constitutes 'scientific realism'
is overly narrow. For example, while he is correct in pointing out that the
'realism' that Feyerabend advocated in many of his major publications
was 'a *normative* claim about how best to set the aims of science' (p. 188),
he speaks in haste when he declares that such a position cannot be
considered a scientific realist one. Consider that van Fraassen (as quoted
earlier), one of the leading contemporary commentators in the scientific
realism debate, unequivocally defines scientific realism as something to
do with the aims of science.

Tambolo (2014, p. 205) is correct to note that while Feyerabend often
attacked what he called 'realism', his critique 'does not apply to all versions
of scientific realism'. The important point that we can all agree on is that
Feyerabend was fundamentally opposed to some of the attitudes exhibited
by many philosophers who have called themselves 'scientific realists'.
Given that, the interesting question to ask is: what sense can we make of
Feyerabend's realism, and how did that position develop as Feyerabend's
thinking changed over the years? I concur with Matthew J. Brown (2016)
that a pluralist kind of realism, which he calls 'abundant realism', can be
discerned in Feyerabend's late writings, especially *The Conquest of
Abundance* (1999). But, I would also not agree wholly with the view,
expressed for example by Jamie Shaw (2018a, p. 35), that the realism

defended by the later Feyerabend is different from his earlier realism (which Shaw locates in the period from 1951 to 1981). The later Feyerabend's objections to what some philosophers called 'realism' do not amount to, and were not accompanied by, a renunciation of his own earlier stance that he had called 'realism'.

I want to show that Feyerabend's realism only became *stronger* in his later works, at least in one sense: as mentioned previously, his earlier position could be taken as a quasi-realist one: 'we treat scientific theories *as if* they described the world without being committed to the view that theories succeed in doing so' (Shaw 2018a, p. 35). In his later view, as I will explain further, there is no 'as if' involved: reality *is* what good theories describe. This idea also strengthens Feyerabend's pluralism: what he now advocates is not an ultimately frustrating proliferation of provisional 'as if' theories, most of which will fade away, but a proliferation in which each theory, each way of life, helps create a new and different reality, not to be discarded when the correct one is found. Calling this move a 'retreat' from realism does not square with the spirit of Feyerabend's thought; it is only a perspective forced by the presumption that realism cannot be truly pluralist. Oberheim (2006, section 6.3) is correct to note the increasing neo-Kantian tendency in Feyerabend's thinking, but seeing that tendency as anti-realist makes Oberheim undervalue in Feyerabend's thinking what we may fairly call realism, and its continuity. It is true that there was a more semantic and less metaphysical emphasis in his earlier statements, but I do not think that this shift of emphasis amounts to a whole new position; on the contrary, the later *addition* of a metaphysical dimension to Feyerabend's realism only made it more complete and coherent.

2.4.2   Feyerabend's Pragmatist–Pluralist Notion of Reality

I think that already in the 1960s Feyerabend started to make significant moves in the development of his *pluralist realism*. By the time he penned the 'Outline of a Pluralistic Theory of Knowledge and Action' in 1968, the pluralist aspect of his realism began to shine through clearly. Once again defending his 'principle of proliferation', Feyerabend noted that new theories could reveal new facts: 'There may exist facts that endanger [the current theory] *T* but that can be revealed with the help of alternatives only'. If that happens, the alternative theory has 'not just accentuated an already existing difficulty [for *T*]; it has actually created it' (Feyerabend 1968/1999, p. 108). Here one can hear an echo of Thomas Kuhn saying that we have to learn to make sense of how after a scientific revolution scientists would seem to inhabit a different empirical world. In addition, Feyerabend spoke of 'the metaphysical components of observation', which could also be replaced by scientific theories (pp. 109–110). The obvious epistemological dimension of proliferation began to show metaphysical implications. By 1981, he used the phrase 'pluralistic realism' explicitly, and spoke of alternative 'cosmologies and forms of life' (Feyerabend 1981a, xi).

In grappling with Feyerabend's pluralist realism, it is crucial that we understand what Feyerabend meant by 'reality' as he tried to spell that out in the set of texts collected as the posthumous volume *The Conquest of Abundance*. This volume consists of an unfinished manuscript of that title, together with reprints of related essays that were previously published.[6] Among the latter are a few papers that lay out quite clearly Feyerabend's later conception of reality, one that Preston finds beyond the pale: 'One of the most remarkable things about his last work is that he commits himself, albeit tentatively, to a new metaphysical picture of the world, a clear rival to the picture of mind-independent reality that undergirds scientific realism. The replacement is a metaphysic best characterised as "social-constructivist"' (Preston 2000, p. 94). Brown (2016, section 4) gives an excellent account of Feyerabend's later metaphysics, and my own reading of Feyerabend's texts is buttressed by Brown's exegesis.

Feyerabend's paper titled 'Realism', published in 1994, the year of his death,[7] is a bewildering piece; his long discourse on Achilles and honour does not clearly spell out anything about realism. However, the later parts of the paper are clear enough. Feyerabend is not against realism, but against monism and absolutism. He criticises the 'scientific realists' for their hegemonic view of modern science, about which he had already expressed clear disdain in *Against Method* and other writings in the 1970s. A key passage in the 1994 paper is his discussion of the Parmenidean 'realist' argument that change is impossible, and its modern descendants resulting in notions such as the 'block universe'. He is rightly incredulous about these arguments, but his real point is not that such a view is wrong. What he finds really objectionable is the insistence that it is the only correct view.

Feyerabend reminds us that there were various ways of conceiving reality advanced by the ancient Greeks: 'Thus we can say that at the time in question (fifth to fourth century B.C.) there existed at least three different ways of establishing what is real: one could 'follow the argument'

---

[6] As the published papers included in the volume date from 1989 to 1995, I see the development of Feyerabend's thoughts as a fairly continuous one: 1989 was only eight years after the first two volumes of his *Philosophical Papers* were published, and just two years after *Farewell to Reason*.

[7] The opening chapter of *The Conquest of Abundance* closely follows the content of this paper.

[Parmenides]; one could 'follow experience' [Leucippus]; and one could choose what played an important role in the kind of life one wanted to lead [Aristotle]. Correspondingly there existed *three notions of reality* which differed not so much because research had as yet failed to eliminate falsehoods but because there were different ideas as to what constituted research.' So 'it seems that the "problem of reality" has many solutions' (1994/1999, p. 190). All of the solutions were legitimate, though it seems clear that Feyerabend's own sympathies were with the Aristotelian conception. Advocates of each conception of reality should by all means try to make it work; it was Feyerabend's optimistic view that just about any conception of reality could be made to work to a degree, and that many conceptions, including some that are rejected by modern science, could be made to work quite well.

For my argument in favour of continuity in Feyerabend's thinking, it is important to note that already in 1981 he had inserted a pluralist view of reality into the Introduction to the first two volumes of his *Philosophical Papers,* to frame his earlier papers reprinted in that collection: '*we decide to regard those things as real which play an important role in the kind of life we prefer.*' This is quite a perfect statement of what Brown (2016, p. 147) identifies as the fourth central plank in Feyerabend's late 'metaphysics of abundance', namely, 'Aristotle's Principle: What is "real" is what plays a role in our valued practices and form of life, what we care about and identify with.' And in the same text of 1981, Feyerabend also already had a clear negative verdict on monist–absolutist realism: 'Realism . . . only reflects the wish of certain groups to have their ideas accepted as the foundations of an entire civilization and even of life itself' (1981a, p. xiii). A clear difficulty in exegesis here, as noted earlier, is that Feyerabend often just said 'realism' in such passages, without qualifying exactly what kind of realism he had in mind. So, as usual, we cannot avoid the task of careful interpretation.

The italicised statement quoted in the previous paragraph cries out for a pragmatist interpretation, and Feyerabend's later writings do deliver such an interpretation. The following is perhaps the clearest passage: 'putting reality where the achievements are, there are different kinds of reality defined by different modes of successful research' (Feyerabend 1994/1999, p. 194). Such pragmatism goes together well with pluralism – not as a matter of logical or conceptual necessity, but as a matter of historical fact. Feyerabend argues that different conceptions of reality have occurred not only at the level of general metaphysical perspectives, but also at the level of concrete scientific detail: 'the different conceptions of reality that occur in the sciences have empirical backing. *This is a historical fact, not a philosophical position,* and it can be supported by a closer look at scientific practice.' And such differences have not disappeared with the development of science: there are many modern scientists who are in a way 'continuing the Aristotelian approach, which demands close contact with experience and objects', while there are others such as Einstein who prefer the method of 'following a plausible idea to the bitter end' (p. 191). Many different scientific fields with different approaches 'have been successful, thus confirming the notions of reality implicit in their theories' (p. 192).

Feyerabend was by no means the only thinker who advocated such a pragmatist pluralist realism. Israel Scheffler (1999) made a plea for 'plurealism' around the same time as the publication of *Conquest of Abundance,* emphasising how naturally realism can embrace pluralism. Around the same time, Roberto Torretti (2000) gave expression to the view that a viable kind of realism would be a pluralist position, referring explicitly back to Hilary Putnam's ideas from his 'internal realist' phase. As Torretti points out, Putnam (1987, p. 17) mused that he should have used the phrase 'pragmatic realism' instead of 'internal realism' to designate his position. Putnam's and Torretti's pragmatist realism had clear pluralist implications, and I am currently at work in developing these implications further.[8]

2.4.3   Feyerabend's Constructivism

It is in line with common misunderstandings of pragmatism that some critics take the later Feyerabend as a 'constructivist' in a pernicious sense of the term: 'Feyerabend's more radical version of postmodernism has anti-realist implications, bearing negatively both on realism about scientific *theories,* and on realism about scientific *entities*' (Preston 2000, p. 89). And Feyerabend certainly gave some cause for such interpretation. For example, he says that humans are 'sculptors of reality'.[9] In contrast to the standard realists who would insist on the mind-independence of reality ('atoms existed long before they were found', and so on), Feyerabend says: 'A better way of telling the story is the following. Scientists . . . used ideas and actions (and, much later, equipment up to and including industrial complexes such as CERN) to *manufacture,* first, metaphysical atoms; then, crude physical atoms; and, finally, complex

---

[8] See Chang (2016) and Chang (2018) for some preliminary forays. I have made a mistake of composing these papers without mentioning Feyerabend. I hope that the present chapter goes some way towards correcting that mistake.

[9] Or even that nature is a 'work of art' (see Preston 2000, p. 94).

systems of elementary particles out of a material that did not contain these elements but could be shaped into them' (1989/1999, p. 144).

However, it is important to note two things. First of all, what we successfully manufacture is real, after the manufacturing is done; this is just as the etymology of the word 'fact' indicates something that has been (successfully) made. In that sense, there is no conflict between constructivism and realism. Secondly and more importantly Feyerabend, like the pragmatists, is very clear that we cannot construct just any kind of reality we might like: 'I do not assert that any combined causal-semantic action will lead to a well-articulated and livable world. The material humans (and, for that matter, also dogs and monkeys) face must be approached in the right way. It offers resistance; some constructions (some incipient cultures—cargo cults, for example) find no point of attack in it and simply collapse' (p. 145). As Tambolo (2014, section 3) expresses the point: the 'pliability thesis' upheld by the later Feyerabend is clearly tempered by the 'resistance thesis'. Again, Feyerabend had been anticipated by the pragmatists: as William James put it (1907/1978, p. 106), 'Experience, as we know, has ways of *boiling over*, and making us correct our present formulas'.

2.4.4   The Inscrutability of Being

But what exactly is the 'material' that he speaks of in both of the statements just quoted? In some places it seems that what Feyerabend means by the 'material' is the 'world itself' or 'Reality' with a capital 'R'. Preston (2000, pp. 94–97) rightly notes that there is something Kantian about such a view, expressing disapproval. If so, here language (Feyerabend's or anyone's) fails us, and there is no convenient answer to the question. The ultimate Reality is completely inscrutable, like Kantian noumena, as it does not fall under any concepts. How Reality interacts with what we do so that tangible realities are produced, is also inscrutable. About this process, Feyerabend says that we can try to give an account, but any such account would be 'from the inside' of a particular approach that we adopt: 'We can tell many interesting *stories*. We cannot explain, however, how the chosen approach is related to the world and why it is successful, in terms of the world. This would mean knowing the results of all possible approaches' (Feyerabend 1989/1999, p. 145). This view is repeated five years later: 'describing a response and not Being itself, all knowledge about the world now becomes ambiguous and transparent. It points beyond itself to other types of knowledge and, together with them, to an unknown and forever unknowable Basic Reality' (1994/1999, p. 196).

Despite the initial plausibility of the interpretation just given, I think there must be something different, or at least something additional, that Feyerabend meant. If the 'material' is the inscrutable Being, then it makes no sense to talk about how we can *shape* it into anything. It makes more sense to see Feyerabend's 'material' as not the world itself, but the world as we have it according to some inherited account. It is important not to equate this 'material vs. product' distinction with the usual 'appearance vs. reality' distinction, which Feyerabend did not like so much. When the standard realists talk about appearance and reality, they do mean that the reality hidden behind appearances can be known, or at least got at in some partial way. For Feyerabend, appearances *are* realities, and the Being that may be conceived to be 'behind' the apparent realities is ineffable, inscrutable.[10] In this context, it is interesting to note Feyerabend's commentary on Parmenides (p. 188): 'This was the first, the clearest and most radical separation of domains which later were called reality and appearance and, with it, the first and most radical defence of a realist position. It was also the first theory of knowledge. Those who are ready to make fun of Parmenides should consider that large parts of modern science are bowdlerised versions of his result'. Consider modern scientists' attachment to conservation laws, and their surprising willingness to commit to the 'block universe' picture in which the future is fixed in the same way as the past.

One additional clarification is necessary here, to make something approaching a well-rounded picture of the later Feyerabend's metaphysics. This concerns one rare place where I think Brown's excellent exposition needs a slight correction. Brown (2016, p. 149) states: 'when Feyerabend asserts that Being is "ineffable and unknowable," he is *not* making the transcendental idealist metaphysical claim ... that we lack any access whatsoever to Being, etc. Instead ... the ineffability and unknowability of the world follows from its abundance. The complex, overlapping, malleable nature of Being's structure make[s] it impossible to capture in a single formulation.'[11] In speaking of 'Being's structure' being such and such, I think Brown falls into the trap of talking about 'the unknowable ... as if we know what it is like', against which he warns us (ibid.). No, abundance is not the same thing as ineffability. 'Being' or 'Basic Reality', whatever that is, is ineffable, indescribable, unknowable. What is abundant is the richness of experience, and all the different ways in which people

---

[10] See Ian Kidd's (2012) instructive discussion of the link between this notion of ineffability and epistemic pluralism.

[11] In this quotation, I have suppressed in the ellipses the parts of Brown's passage which I think do get things exactly right, to highlight only the parts that need revision.

have known and made sense of experience. The 'conquest' of that abundance can only be managed by the human collective in a pluralist way.

2.5   Proliferation and Human Flourishing

In closing, it would be appropriate to make a brief consideration of Feyerabend's deeper motivations. In relation to his earlier works, testability and 'anti-rationalism' were identified as the drivers of his thinking about realism. At least from *Against Method* onwards, Feyerabend made it clear that his sights were set on no less than human flourishing in a general sense. And Feyerabend's inclination concerning human flourishing was strongly towards diversity and freedom. It is not an accident that he looked to John Stuart Mill's *On Liberty* for his inspiration and arguments for pluralism (e.g. Feyerabend 1981c, pp. 139–141; Lloyd 1997). Rejecting the Platonic ideal of the Good as an inoperative one, Feyerabend held that the good was to be sought and worked out by each individual and each community of human beings. He quoted Aristotle's *Nichomachean Ethics* on this point: 'Even if there existed a Good that is one and can be predicated generally or that exists separately and in and for itself, it would be clear that such a Good can neither be produced nor acquired by human beings' (Feyerabend 1994/1999, p. 189). Where human freedom and diversity meet the abundance of nature, the only reasonable attitude is tolerance and openness.

From Feyerabend's view of life, it is obvious that the forcing of one mode of inquiry and one mode of life on other people must be done with the greatest reluctance, and that it is much more rarely necessary than people imagine. This view he laid out quite clearly at the end of *Against Method*, and more extensively in *Science in a Free Society*. In 'Consolation for the Specialist' he declared: 'Progress has always been achieved by probing well-entrenched and well-founded forms of life with unpopular and unfounded values. This is how man gradually freed himself from fear and from the tyranny of unexamined system' (Feyerabend 1970a, pp. 209–210). He was clearly against the misuse of philosophical notions often associated with 'realism' as oppressive devices: 'One of my motives for writing *Against Method* was to free people from the tyranny of philosophical obfuscations and abstract concepts such as "truth", "reality" or "objectivity", which narrow people's vision and ways of being in the world' (Feyerabend 1999, p. viii, quoted from *Killing Time*). Feyerabend also refused to regard science as a body of knowledge isolated from the rest of life. As the years went by, he increasingly and adamantly viewed science as a set of human practices, and inquiry as a natural and indispensable part of human life. Here again there is remarkable affinity between Feyerabend's thinking and pragmatist philosophy, especially some key ideas of John Dewey.

In closing, I must note that I am giving an interpretation of Feyerabend that looks very close to some views that I have been developing myself (e.g. Chang 2016, 2018). Does this signal a bias in my interpretation? Yes, but I think it is a helpful bias. Feyerabend has often been misunderstood because many commentators ultimately regard his views as absurd, even when they study them closely. *That* bias results in a rendition of Feyerabend's views in a more absurd form than necessary, leading to hasty dismissal. My bias comes from sympathy, which makes me inclined to formulate his views in the most defensible way. Not only is this in line with the old 'principle of charity', but also especially in the case of Feyerabend, it serves as a useful antidote to the general atmosphere of hostility and misunderstanding. And my interpretation has been tested empirically in the same sort of way in which many enemies of Feyerabend would insist scientific theories should be tested, namely by independent confirmation. Initially, I was inspired by Feyerabend's pluralism, with no knowledge of his commentary on realism; my own Feyerabend-inspired pluralism developed into a kind of realism, and found a congenial home in pragmatism; and then when I came to read Feyerabend's comments on realism, I found them to be resonant with my own views, in line with what I think a Feyerabendian thinker should think, and what I thought Feyerabend *would have said* about realism. So, if my own interpretive lens seems to be refracting Feyerabend's views too strongly, that is not quite how the situation is. Rather, my lens is itself Feyerabendian, well suited for viewing Feyerabend.[12]

[12] Acknowledgements: I would like to thank Jamie Shaw for his invaluable and generous guidance, without which I would never have been able to write this chapter. I also thank Karim Bschir and Paul Hoyningen-Huene for their incredible encouragement in my study of Feyerabend.

CHAPTER 3

Feyerabend's General Theory of Scientific Change

Hakob Barseghyan

3.1 Introduction

Why did Feyerabend never detail a general account of the patterns by which various elements of scientific theorising (theories, methods, etc.) change over time? In other words, why did he never author a general theory of scientific change? To anyone even remotely familiar with the works of Feyerabend, this question may sound absurdly out of place. After all, Feyerabend spent his whole career arguing against the alleged rationality of science and fervently opposing the idea of a fixed and universal method of theory evaluation. With his assertion that 'science is not one thing, it is many' (Feyerabend 1992, p. 6; see also Feyerabend 2011, p. 56), he is nothing short of a godfather of the disunity of science movement that revels in its explicitly anti-theoretical approach to the process of scientific change (see Lloyd 1996, p. 261). Surely, Feyerabend would be the last person to search for any general patterns of scientific change!

While it is true that creating general theories was never among the wide array of Feyerabend's intentions, a close examination of Feyerabend's writings reveals that he foreshadowed – perhaps accidentally – several general *patterns* of scientific change that are currently captured by the scientonomic laws of scientific change.[1] Thus, Feyerabend knew that theories are normally expected to satisfy the methods of the respective

[1] The four axioms and twenty odd theorems of this general descriptive theory explain many aspects of the process of scientific change. Among other things, the theory explains how theories and methods of their evaluation change through time in a law-governed fashion (Barseghyan 2015; Sebastien 2016; Patton et al. 2017; Rawleigh 2018; Barseghyan 2018). A special workflow has been implemented to communally advance the theory in a piecemeal and transparent fashion (Shaw & Barseghyan 2019). The current state of the theory, the history of its recent developments and the new workflow are presented in the *Encyclopedia of Scientonomy* (www.scientowiki.com).

community at the time of the assessment, that the 'methods of selling depend on the audience' (Feyerabend 1973, p. 115). He also knew that these methods are normally shaped by the community's accepted theories; that is, the accepted theories 'provide standards to judge other *theories*' (Feyerabend 2010, p. 243). One finds many historical illustrations of these patterns in Feyerabend's analysis of historical episodes.

Yet, paradoxically, Feyerabend himself never derived a conclusion by constructing a general theory of scientific change. Why did he fail to take that seemingly obvious step? If he knew that that theories are normally expected to meet the standards of the respective community at the respective time, and if he knew that these standards are usually shaped by the community's accepted theories, why did he never accept that there are, after all, some general patterns of scientific change? The goal of this chapter is to provide an answer to this conundrum.

Two major factors, I argue, contributed to Feyerabend's failure to see the patterns that he himself so astutely described on so many occasions. The first and most obvious is his *particularist axiology*, that is, his aim at locating the exceptional and quirky, rather than the general and orderly. The second is Feyerabend's failure to draw or fully appreciate two important historiographic distinctions. Specifically, he questioned the distinction between *acceptance* and *pursuit* as different epistemic stances that can be taken towards theories. In addition, while differentiating between *openly prescribed* methods of theory evaluation and methods *actually employed* in theory evaluation, he didn't fully appreciate the role of the latter. His focus on pursuit and openly stated methodological dicta, rather than acceptance, and actually employed methods only reinforced his particularist axiology and prevented him from spotting the general patterns of scientific change revealed in his thorough historical analyses.

3.2   Patterns of Scientific Change

Feyerabend's main objective, during the 1970s, was to debunk the idea of one universal and unchangeable method of science and to show that the methods (standards, criteria) of theory evaluation are as changeable as theories themselves.[2] His goal was to show that 'there is not a single rule, however plausible, and however firmly grounded in epistemology, that is

---

[2] This conflicts with notable interpretations of Feyerabend's anarchism, such as Farrell (2003), who claims that the process of theory evaluation is unmethodical. For the purpose of this chapter, I will rely on the interpretation provided by Shaw (2017, 2018).

not violated at some time or other' (Feyerabend 2010, p. 7). By doing so, he greatly contributed to the rise of the disunity of science movement and the idea that there are no general patterns of scientific change that hold for all communities at all times. In fact, his magnum opus is one of the theoretical foundations on which the whole disunity of science movement is based (Dupré 1993; Hacking 1996; Chang 2012). Yet, in his analyses of historical episodes, Feyerabend revealed – perhaps accidentally – several general patterns of scientific change, specifically those concerning changes in theories and methods of their evaluation.

It's safe to say that any viable descriptive theory of scientific change should explain how both theories and methods of their evaluation change through time (Barseghyan 2015, pp. xiii, 3–11). Indeed, once we accept that methods of theory evaluation change through time, it is no longer viable to view all transitions in science as mere changes in theories, that is, as a series of transitions from one theory to the next where subsequent theories are deemed better by some universal standards of a fixed scientific method.[3] Since it is currently well known that there is no such thing as a fixed and universal method of science, a general theory of scientific change should also explain how methods themselves change through time. Thus, any successful theory of scientific change will have to explain the mechanism of theory change *and* the mechanism of method change, among many other things.

Having argued that the standards of theory evaluation are changeable, local and contextual, Feyerabend essentially described one of the key general patterns of scientific change – the idea that theories become accepted only when they meet the requirements of the community at the time of the assessment (Patton et al. 2017). Consider, for example, his treatment of the experts' reaction to Galileo's ideas (Feyerabend 2010, pp. 128–129):

> Experts (*qualificatores*) were ordered to give an opinion about two state-ments which contained a more or less correct account of the Copernican doctrine . . . [T]he experts declared the doctrine to be 'foolish and absurd in philosophy' or, to use modem terms, they declared it to be unscientific. This judgement was made without reference to the faith, or to Church doctrine, but was based exclusively on the scientific situation of the time. It was shared by many outstanding scientists (Tycho Brahe having been one of them) – *and it was correct* when based on the facts, the theories and the standards of the time.

---

[3] Classical theories of scientific change of Popper (Popper 1959, 1963), Lakatos (Lakatos 1970; Lakatos and Zahar 1976), and early-Laudan (Laudan 1977) were based on the assumption that, while theories change through time, there are fixed methodological standards of theory evaluation.

While Feyerabend's main goal here is to show that Galileo's theory was not acceptable according to the standards of the time, he also coincidentally revealed a general pattern of theory acceptance – they become accepted only when they meet the requirements of the method employed at the time. According to Feyerabend, there was no violation of the *method* employed at the time, but only a violation of twentieth-century methodologies of positivists, empiricists, Popper and Lakatos (pp. 145–147).[4] Similarly, Feyerabend emphasises that 'the methods of selling depend on the audience' (Feyerabend 1973/1999, p. 115). Thus, while opposing universal standards, he also inadvertently reveals that theories are normally accepted when they meet the standards of the respective community at the respective time. This is the gist of the scientonomic second law of scientific change (Patton et al. 2017).

Ironically, while arguing for the absence of fixed and universal methodological standards, Feyerabend also provided an exposition of *how* these standards change. Despite his official stance, which he first labelled as *dadaist* and later as *anarchist* (Feyerabend 1973/1999, pp. 114–115), Feyerabend clearly admits that there is 'the way in which science . . . revises its "standards"' (Feyerabend 2010, p. 243). Chapter 18 of *Against Method* alone provides several examples illustrating how accepted theories shape the standards of theory evaluation. Consider the following passage (p. 243):

> In physics theories are used both as descriptions of facts and as standards of speculation and factual accuracy. Measuring instruments are constructed in accordance with laws and their readings are tested under the assumption that these laws are correct. In a similar way theories giving rise to physical principles provide standards to judge other theories by: theories that are relativistically invariant are better than theories that are not. Such standards are of course not untouchable. The standard of relativistic invariance, for example, may be removed when one discovers that the theory of relativity has serious shortcomings.

Thus, according to Feyerabend, the acceptance of general relativity gives rise to the requirement that a new physical theory that doesn't aim at replacing general relativity should be relativistically invariant to be acceptable. Similarly, in the following passage, he argues that the acceptance of the idea that the world is qualitatively and quantitatively infinite gave rise to the requirement of excess empirical content (p. 243):

---

[4] In addition, Feyerabend also foreshadowed the *contextual appraisal theorem*, that theory assessment is contextual and depends on the accepted theories and employed methods of the time, currently accepted in scientonomy (Barseghyan 2015, pp. 183–196).

The idea that nature is infinitely rich both qualitatively and quantitatively leads to the desire to make new discoveries and thus to a principle of content increase which gives us another standard to judge theories by: theories that have excess content over what is already known are preferable to theories that have not. Again the standard is not untouchable. It is in trouble the moment we discover that we inhabit a finite world. The discovery is prepared by the development of 'Aristotelian' theories, which refrain from going beyond a given set of properties – it is again prepared by research that violates the standard.

Feyerabend shows that if scientists were to accept that the world is quantitatively and qualitatively finite, they would be reluctant to accept theories that introduce unfamiliar properties (pp. 244–245). Here is another example from Feyerabend illustrating how our accepted beliefs shape or criteria of theory evaluation (p. 245):

> The idea that information concerning the external world travels undisturbed via the senses into the mind leads to the standard that all knowledge must be checked by observation: theories that agree with observation are preferable to theories that do not. This simple standard is in need of replacement the moment we discover that sensory information is distorted in many ways. We make the discovery when developing theories that conflict with observation and finding that they excel in many other respects.

This same pattern can also be found in Feyerabend's early works. For instance, speaking of Bohr's complementarity principle, he shows how the discovery of the wave–particle dualism led to the replacement of the 'the classical ideal of explanation', that is the classical method of theory evaluation, according to which an acceptable theory 'must be universal, i.e., it must be of a form which allows us to say what light is rather than what light appears to be under various conditions' (Feyerabend 1958c, p. 78):

> For Bohr the dual aspect of light and matter is not the deplorable consequence of the absence of a satisfactory theory, but a fundamental feature of the microscopic level. For him the existence of this feature indicates that we have to revise, not only the classical physical theories of light and matter, but also the classical ideal of explanation.

Thus, Bohr rejects the classical method of theory evaluation and proposes the complementarity principle, which suggests that an acceptable physical theory may not necessarily provide a universal description, but may elucidate the object under study from diverse points of view.

Nowadays, we can easily take these examples as solid illustrations of the closely parallel scientonomic third law of scientific change which states that

'A method becomes employed only when it is deducible from some subset of other employed methods and accepted theories of the time' (Sebastien 2016, p. 4).[5]

Thus, it is clear that Feyerabend accepted that theories are evaluated by the methodological standards of the time and that these standards somehow follow from other accepted theories. With these two key ingredients in place, he could have noticed that theories and methods change in a sort of *iterative* way, where new theories become accepted when they meet the current standards and, once accepted, often lead to changes in these standards – an idea that is currently captured both in Chang's conception of epistemic iterations (Chang 2004) and the scientonomic laws (Barseghyan 2015; Sebastien 2016; Patton et al. 2017). Yet, Feyerabend didn't take this seemingly natural step.

A question arises here: if Feyerabend anticipated the key patterns of scientific change, how could he fail to admit that both theories and methods change in a law-governed fashion after all? Indeed, if Feyerabend saw that theories are normally accepted when they meet the requirements of the methods employed at the time, and if he saw that the requirements themselves are shaped by previously accepted theories, then why did it never occur to him to take a seemingly obvious step and build a general theory of scientific change? How could he possibly skip making the proverbial soup after he personally obtained most of its required ingredients?

I believe there are two interconnected factors here: Feyerabend's aim at locating exceptions (his *axiological particularism*) and his failure to draw or fully appreciate the consequences of some important historiographic distinctions. In what follows, I will discuss each of these factors in turn.

3.3 Axiological Particularism

Having all the required ingredients of a soup doesn't necessarily result in making the soup; one also needs an *intention* to make the soup. It is evident that Feyerabend wouldn't create a general theory of scientific change, even if he had all the required ingredients available to him, for his aim was not generalist, but *particularist*. Here is an indicative quote from *Feyerabend's Last Letter* (Feyerabend 2010, p. xv): 'What I want to do is

---

[5] It follows from Feyerabend's examples that theories of *different* types can shape the methods of theory evaluation. For Feyerabend, changes in method are not necessarily a result of changes in scientific theories strictly speaking, as they 'are often tied to metaphysical beliefs' (Feyerabend 2010, p. 253).

to change your attitude. I want you to sense chaos where at first you noticed an orderly arrangement of well behaved things and processes.' Feyerabend's axiological particularism is best seen in opposition to Lakatos's axiological generalism. It might be argued that many of the specific points of contention between the two stemmed from the two fundamentally different axiological starting points: where Lakatos aimed to see general patterns, Feyerabend looked for ambiguities. Where Lakatos sought order, Feyerabend sought chaos. Where Lakatos found rules, Feyerabend found exceptions.[6]

The two axiological viewpoints face drastically different challenges. While the challenge of the particularist is to deal with the existence of regularities, the challenge of the generalist is to deal with the existence of anomalies (p. xvi). Thus, both Lakatos and Feyerabend clearly saw the proverbial 'ocean of anomalies' – not only in how nature defied the tenets of even the most successful theories, but also in how the process of scientific change itself defies the rules of even the most sophisticated methodology (Lakatos 1970, pp. 6, 50, 53; Feyerabend 2010, p. 33). Yet, they differed in their respective reactions to these historical anomalies. According to Lakatos, the historical anomalies – that is, discrepancies between the methodological dicta and the actual course of scientific change – are to be relegated to the footnotes (Lakatos 1971, p. 120):

> One way to indicate discrepancies between history and its rational recon-struction is to relate the internal history *in the text*, and indicate *in the footnotes* how actual history 'misbehaved' in the light of its rational reconstruction.

In contrast, Feyerabend revels in the chaotic, messy and disunited char-acter of scientific change and ridicules any attempts to shoehorn the true complexity of the process into the confines of one's favourite methodology (Feyerabend 2010, p. 11):

> [T]he history of science will be as complex, chaotic, full of mistakes, and entertaining as the ideas it contains, and these ideas in turn will be as complex, chaotic, full of mistakes, and entertaining as are the minds of those who invented them. Conversely, a little brainwashing will go a long way in making the history of science duller, simpler, more uniform, more 'objective' and more easily accessible to treatment by strict and unchange-able rules.

---

[6] Obviously, this is not to imply that Feyerabend didn't occasionally make generalist remarks. Consider, for instance, his insistence that traits he discerned in his analysis of the case of Galileo are not an exception but a rule (Feyerabend 2010, p. 40 fn. 20).

Feyerabend, as far as I can tell, was convinced that whatever purported regularities there might be, they will all crumble upon closer attention. In 'Experts in a Free Society', he writes (Feyerabend 1970b, p. 123):

> A scientist finds himself in a complex historical situation. There are observations, attitudes, instruments, ideologies, prejudices, errors, and he is supposed to improve theories and change minds under the highly individualized circumstances created by the interplay of all these factors. Instruments as well as people must be coaxed into giving the proper response, taking into consideration that no two individuals (no two scientists; no two pieces of apparatus; no two situations) are ever exactly alike and that procedures should therefore be allowed to vary also.

The same axiological position is evident in many other passages. For instance (Feyerabend 1972, p. 171):

> [S]cience should be considered as a complex and a highly heterogeneous *historical process* in which vague and unrelated anticipations of future ideologies are developed side by side with highly sophisticated theoretical systems and petrified forms of thinking.

Similarly, in *Science in a Free Society* he writes (Feyerabend 1978b, p. 211):

> [L]ogical rules are too simpleminded to be able to reflect the complex structures and movements of scientific change.

It is Feyerabend's deep conviction that any general claims about science can be retained only as 'a verbal ornament, as a memorial to happier times when it was still thought possible to run a complex and often catastrophic business like science by following a few simple and 'rational' rules' (Feyerabend 1970a, p. 149). Thus, it was never Feyerabend's goal to discover any regularities in the process of scientific change. Instead, his main aim was to show that the twentieth-century methodologies had little to do with how science actually works (Feyerabend 2010, pp. 145–147) and that 'the belief in a unique set of standards that has always led to success and will always lead to success is nothing but a chimera' (p. 164). That is why, having provided so many vivid examples of how methods are shaped by accepted theories, Feyerabend never bothered to present the mechanism of this shaping as a *general pattern*, let alone build a general theory of scientific change. In contrast, Lakatos, who had spent most of his professional career vehemently opposing Feyerabend's idea of the changeability of scientific methods, announced his intention to build such a general theory the moment he became convinced that methods change after all. In one of his final letters to Feyerabend, dated 10 January 1974, he conceded that methods of theory evaluation do change through time. His new book titled *Changing Logic of Scientific Discovery* was supposed to explain how newly accepted theories bring about changes in methods of theory evaluation (Motterlini 1999, p. 355; see also Motterlini 2002). Alas, this project was never to be completed by Lakatos, as he died only three weeks after announcing it.[7]

In brief, it seems reasonable to conclude that the key to resolving our conundrum is Feyerabend's axiological particularism. Yet, a simple solution along the lines of 'he saw the trees but failed to notice the forest' doesn't provide the full story.

3.4   Historiographic Distinctions

A more focussed reading of Feyerabend reveals that in addition to his lack of motivation to notice general patterns, he also didn't draw or fully appreciate the significance of two crucial historiographic distinctions. Specifically, the distinctions between *acceptance* and *pursuit* as distinct epistemic stances one can take towards theories and between scientists' *openly prescribed methodological standards* and their *actual methodological expectations*.

The distinction between *acceptance* and *pursuit* has deep historical roots and this is not an opportune place to delve into it. Suffice it to say that it can be traced as far back as Laudan's *Progress and its Problems*, where he distinguishes between 'the context of acceptance' and 'the context of pursuit' (Laudan 1977, pp. 108–114). Among others who emphasise the importance of this distinction are Wykstra (1980, p. 216), Whitt (1990), Achinstein (1993), Franklin (1993), Brown (2001, pp. 90–91) and Patton (2012). The distinction also plays an important part in the scientonomic ontology of epistemic stances, where *theory acceptance* is currently defined as follows (Barseghyan 2018, p. 29):

> A theory is said to be accepted by an epistemic agent if it is taken as the best available answer to its respective question.

It is in this sense that medieval scholastic natural philosophers accepted the Aristotelian theory of four elements, nineteenth-century physicists accepted Newtonian physics and contemporary biologists accept the

---

[7] It took another thirty years before the iterative nature of the process of changes in theories and methods was brought to focus (Chang 2004; Elliott 2012) and yet another decade for the patterns of theory acceptance and method employment to be explicitly articulated as the *second* and *third laws* of scientific change (Barseghyan 2015).

theory of evolution. Acceptance in this technical sense is distinguished from *theory pursuit*, which is defined as follows (Barseghyan 2015, p. 31):

> A theory is said to be pursued if it is considered worthy of further development.

The stance of pursuit is independent of acceptance: while it is possible to accept and pursue a theory at the same time, it is possible to accept a theory without pursuing it or pursue a theory without accepting it. The latter is best illustrated by superstring theories, some of which are considered pursuit-worthy, but not currently accepted by the physics community at large.[8]

As Jamie Shaw and I have shown elsewhere, the lack of a systematic taxonomy of epistemic stances was partially responsible for some of the misunderstandings between Feyerabend and Lakatos (Barseghyan & Shaw 2017). This was one of the key points of misunderstanding between him and Lakatos, who tried to explain that methodological rules are not meant to restrict what theories scientists can work on; after all, any research programme can be worth pursuing, no matter how degenerating it might be. The task of a methodology is to tell us which programme is better. To use more contemporary language, Lakatos didn't want to limit what theories one can pursue, while offering rules for evaluating which theory one should accept. Feyerabend, in contrast, seems to be fully concerned with not imposing any limitations on theory pursuit, when he says 'It is rational to pursue a research programme on its degenerating branch even after it has been overtaken by its rival' (Feyerabend 1973/1999, p. 116). Yet, he doesn't quite see the difference between the two stances and ends up claiming that there is no essential difference between his position and that of Lakatos (p. 116): 'There is, therefore, no "rational" difference between the methodology of Lakatos and the "anything goes" of the anarchist. But there is considerable difference in *rhetorics*'. Not only did Feyerabend not use the distinction between acceptance and pursuit, he also actively denounced it in his review of Laudan's *Progress and Its Problems* (Feyerabend 1981d, p. 67):

> [N]or is it clear that we are dealing with a difference between acceptance and pursuit rather than a difference between different forms of pursuit. Some pharmacologists may of course say 'this is it!' and stop looking for side effects but it would be more than a little absurd to honour such an attitude by creating a special epistemological category; and if the category exists

---

[8] For a more through discussion of the distinction, see Barseghyan (2015, pp. 30–42) and Barseghyan and Shaw (2017, pp. 3–6).

despite its absurdity then it is wise, in the interest of human welfare, to stay on the side of pursuit and to warn patients of doctors who have moved over to acceptance. I conclude that the distinction between acceptance and pursuit may characterise *special* cases but it would be either vacuous or, if not vacuous, unwise to make it a basis of general rules for the evaluation of research traditions.

His sceptical attitude towards this distinction, among other things, led to an equivocal Feyerabendian idea that the process of scientific change is often an irrational, anarchist enterprise (Feyerabend 2010, p. 115):

> Now, what our historical examples seem to show is this: there are situations when our most liberal judgements and our most liberal rules would have eliminated a point of view which we regard today as essential for science, and would not have permitted it to prevail – and such situations occur quite frequently. The ideas survived and they *now* are said to be in agreement with reason. They survived because prejudice, passion, conceit, errors, sheer pigheadedness, in short because all the elements that characterize the con-text of discovery, *opposed* the dictates of reason and *because these irrational elements were permitted to have their way.* To express it differently: *Copernicanism and other 'rational' views exist today only because reason was overruled at some time in their past.*

It is safe to say that a majority, if not all, such Feyerabendian examples of 'irrational survival' concern theory *pursuit* rather than acceptance (see Shaw 2020). What his analyses show is that often scientists dare to pursue theories in violation of their contemporary methodological dicta. Lakatos would be the last person to disagree with him on this. But because Feyerabend lacks any discernible distinction between pursuit and acceptance, he can be – and has been – easily understood as saying that the process of transitions from one accepted theory to the next sometimes goes in violation of the methods of the time. This latter claim is in striking contrast with what Feyerabend says about theory acceptance in other places. For example, when articulating his version of anarchism, Feyerabend states clearly that while he accepts that 'both absolute rules and context-dependent rules have their limits', he disagrees 'that all rules and standards are worthless and should be given up' (Feyerabend 1977, p. 368 fn. 1).

Feyerabend's principle of proliferation,[9] when read with the acceptance/pursuit distinction in mind, says either that any theory is worth accepting, or that any theory is worthy pursuing, or both (Feyerabend 1965b, pp. 105–106):

---

[9] See Shaw (2017, pp. 4–5) for a detailed description of the principle of proliferation.

Invent, and elaborate theories which are inconsistent with the accepted point of view, even if the latter should happen to be highly confirmed and generally accepted. Any methodology which adopts the principle will be called a pluralistic methodology.

If this principle were meant to say that any theory can be *pursuit-worthy*, rationalists like Lakatos would have no issue with it. But if the principle were meant to say that any theory is *acceptable*, then this would go against Feyerabend's own idea that theory acceptance depends crucially on the local standards of the respective community. Thus, we can safely assume that Feyerabend's main interest was to show that theory pursuit is not a rule-governed activity. We see this in many of his passages, where he talks of scientists *pursuing* theories. Here is a typical passage (Feyerabend 2010, p. 251):

But is it not true that scientists proceed in a methodical way, avoid accidents and pay attention to observation and experiment? Not always. Some scientists propose theories and calculate cases which have little or no connection with reality.

Clearly, what he means here is that scientists often pursue theories in violation of methodological rules. Likewise, in his criticism of Schrödinger's requirement that physical theories should be visualisable, Feyerabend provides historical examples of theories being pursued without being initially visualisable but only becoming visualisable at some later stage (Feyerabend 1948; for a discussion, see Kuby 2016; Shaw 2020, section 3.1). The structure of his argument reveals his deeply entrenched conviction that methodologies are meant to apply to theory pursuit.

What he tries to show with these and other similar examples is that theory *pursuit* is not really a rule-governed activity and 'important results come from the confluence of achievements produced by separate and often conflicting trends' (Feyerabend 2010, p. 253). In doing so, he aims to oppose primarily the Popperian school (including Lakatos), as he mistakenly believes that the methodological rules of the Popperian school are meant to apply to theory pursuit. And because his focus is on the chaotic nature of theory pursuit, he fails to fully appreciate what he himself made quite clear concerning theory acceptance, that is, theories become accepted when they meet the local standards of the community at the time of evaluation.

The second important distinction is that between scientists' *openly prescribed methodological dicta* and their *actual methodological expectations*. While Feyerabend clearly draws this distinction, he fails to fully exploit it in his historical case studies. This distinction is not new and can be traced to Einstein's advice to focus on what physicists *do* rather than what they *say* they do or should do (Einstein 1934; see also Westfall 1971, p. 41; Wykstra 1980, p. 211; Lugg 1984, pp. 436–438). The distinction is implicit in Lakatos's conception who accepts that methodological standards may or may not coincide with the implicit method of science (Lakatos 1971). It also plays an important role in the current scientonomic theory (Barseghyan 2015, pp. 52–61 Barseghyan and Mirkin 2019). Feyerabend clearly understood that the two are by no means the same. For instance, in his analysis of the history of empiricism, he distinguishes three distinct periods, the second of which is precisely the period of 'schizophrenia' when what is proclaimed is very different from what is practised (Feyerabend 1965a, p. 154):

> The second period is characterized by a kind of schizophrenia. What is propagated and declared to be the basis of all science is a radical empiricism. What is *done* is something different. This difference between the professed philosophy and the actual practice is covered up both by a manner of presentation which makes it appear that the theories are indeed nothing but true reports of fact and by a tradition of history writing whose function has been described as being purely "ritualistic," and which creates the impression that after an initial revolution in the Renaissance, science has been steadily progressing through the accumulation of more and more facts. The developments occurring under this guise constitute one of the most interesting chapters in the history of thought. The false idea is adhered to as closely as possible in words; the actual results deviate radically from it. The crisis of physics brought about by relativity and by the discovery of the quantum of action terminates this period of schizophrenia.

The distinction clearly didn't escape Feyerabend. Yet, it is unclear if Feyerabend fully appreciated the importance of focussing on the *actual* expectations of epistemic agents, rather than their explicit methodological proclamations.

While it is not controversial to claim that both openly prescribed methodological dicta and scientists' actual expectations change, it is important to appreciate that the latter – not the former – plays a key role in theory evaluation. When agents evaluate theories, they do so by the requirements of their employed methods, that is by their actual expectations. It seems that at times Feyerabend comes close to appreciating this, for example when he stresses that 'methods of selling depend on the audience' (Feyerabend 1973/1999, p. 115). Yet, evidently, Feyerabend fails to fully appreciate this, which explains why he doesn't notice that theories become accepted only when they meet the actual expectations of the community of the time. Thus, while he provides many examples of this pattern, having failed to focus on actually employed methods, he mostly emphasises the fact that often theories violate the openly accepted methodologies – a feature of science with which many rationalists would readily agree (see Worrall 1988; 1989).

Since Feyerabend mostly deals with explicit methodological dicta, he has a hard time reconciling the process of theory acceptance 'with familiar principles of theory evaluation' (Feyerabend 2010, p. 135). In many cases, Feyerabend's *method* refers to what scientists say they should be doing not what they actually do (see, for instance, 150, 164). There are many instances where he clearly refers to the standards that are openly accepted and taught (p. 165). This is partly because his major goal is to demonstrate that the actual practice of science has little to do with the popular twentieth-century methodologies, for 'the naive and simple-minded rules which methodologists take as their guide' cannot possibly account for the 'maze of interactions' found in the process of scientific change (Feyerabend 2010, p. 1). Even the methodology of his arch-rival and friend Lakatos, according to Feyerabend, is at best applicable to the science of the last two centuries (Feyerabend 1973, p. 117). For him, there is an 'abyss' between 'the customary methodologies' and the actual historical cases (Feyerabend 1970c, p. 277):

> I now set myself the task of widening this abyss so that the mechanisms which underlie the actual development of knowledge will stand out and be recognized more easily . . . The aim of the present essay is exactly the same: to progress by emphasizing the contrast between the customary methodologies and certain important episodes in the history of thought.

Even when Feyerabend talks about 'the standards of the time' (Feyerabend 2010, p. 129), it remains unclear whether he means the openly stated methodological dicta of the time or the expectations of the community of the time. For instance, when discussing the Galileo case, Feyerabend can be interpreted as claiming that Galileo's theory didn't meet the actual requirements of the time. However, he can also be interpreted as arguing that Galileo's heliocentrism violated the methodological dicta openly prescribed at the time. This vagueness partially stems from the fact that, as far as I can tell, the actual expectations of the early seventeenth-century Aristotelians were in accord with their openly prescribed methodological dicta (Barseghyan 2015, pp. 143–145). However, it also has to do with Feyerabend's not distinguishing between openly prescribed and actually employed standards.

By focussing on theory pursuit and openly prescribed methodologies, Feyerabend naturally ended up fixated on the case of Galileo with its numerous apparent violations of methodological dicta to argue 'that such violations are not accidental events' but 'that they are necessary for progress' (Feyerabend 2010, p. 7). While he clearly knew that seventeenth-century Aristotelians would only accept a theory if it were intuitively plausible (p. 152), he failed to fully appreciate that it is this requirement that Cartesian and Newtonian theories managed to meet to become accepted. Similarly, while he knew that the requirement of intuitive plausibility 'disappeared from methodology the very moment intuition was replaced by experience, and by formal considerations' (p. 7), he didn't see that it was the acceptance of Newton's theory in Britain and Descartes' theory on the continent that eventually triggered this transition in methods (Barseghyan 2015, pp. 143–150, 211–214). Instead, he sought to explain the transition as a change in methods *first* and theories *second*; he linked the transition in methods to changes in cultural preferences – those who prefer technology to harmony with Nature would value empirical precision and would end up with Newtonian rather than Aristotelian standards (Feyerabend 2010, p. 247).

3.5 Conclusion

In short, Feyerabend had neither *all* the necessary ingredients of a general theory of scientific change nor the *intention* to build such a theory. I have shown that the key reason behind Feyerabend's failure to acknowledge the general patterns that he elucidated was his *particularist axiology* – his orientation towards finding exceptions rather than general patterns. While his numerous historical examples seem to suggest that theories become accepted by a community when they meet the community's criteria, and that these criteria are shaped by the community's accepted theories, Feyerabend would clearly deny that these patterns hold *at all times*. Two major conflations contributed to Feyerabend's conviction that any regular pattern of scientific change must necessarily be riddled with exceptions. As a result, he spilled a lot of ink showing how theories often become *pursued* in violation of *openly* prescribed standards, whereas very few of his historical examples can be interpreted as concerning changes in *accepted* theories and *actually employed* methods of theory evaluation. It was Feyerabend's particularist axiology, reinforced by a few crucial conflations that prevented him from noticing the general patterns of scientific change which were within his grasp.[10]

---

[10] Acknowledgements: I would like to thank Jamie Shaw for his generous help with this chapter. This chapter would be impossible without his encyclopedic knowledge of Feyerabend's corpus.

CHAPTER 4

Feyerabend's Theoretical Pluralism
An Investigation of the Epistemic Value of False Theories

K. Brad Wray

4.1 Introduction

Alternative theories play an important role in Paul Feyerabend's conception of methodology. Because facts and theories are inextricably entwined, even apparently false and unsuccessful theories can play a vital constructive role in theory evaluation (see, for example, Feyerabend 1975/1988, p. 33). Competitor theories can draw attention to hitherto unnoticed facts. And some facts are even wholly undetectable without the aid of an alternative theory. In response to the new facts, proponents of competing theories are pressed to either (i) accept the new theory that exposed the new facts or (ii) develop the accepted theory to account for the new facts. This is one of the important ways in which science progresses. Our theoretical understanding of the world is thus enriched by theoretical pluralism.

The radical nature of Feyerabend's view may not be so easy to appreciate for contemporary readers for whom pluralisms of various kinds are commonplace (see, for example, Kellert et al. 2006; Chang 2012; and Brown and Kidd 2016). But, when Feyerabend initially presented this argument, it was widely assumed that the scientists working in a research community are united by their shared commitment to a *single* theory at a given time in a given domain. This is evident not only in the writings of the Logical Positivists, but also in the works of Karl Popper and Thomas Kuhn. Normal science, that is, the sort of research for which scientists are trained, is characterised by a commitment on the part of those working in a specialty to a *single* theory. And revolutionary changes of theory seem to presuppose theoretical monism as well. After all, a revolution has only run its course when a single theory has been widely accepted, and a new normal scientific research tradition is set in motion, one based on the newly accepted theory (see Kuhn 1962/2012; on Kuhn's monism, see Feyerabend 1965/1981, p. 108 fn. 14).

In this chapter, I re-examine the case for theoretical pluralism. Feyerabend emphasises that even obviously false theories can play a constructive epistemic role. False theories can aid scientists in seeing the limitations of the theories they currently accept. Recent defences of theoretical pluralism are quite different from Feyerabend's defence of pluralism. They tend to emphasise the fact that scientists are more likely to find a true theory or one closer to the truth when scientists value pluralism, and a greater range of theories are considered. The more theories that scientists have to consider, the more likely they are to find the right one, so the argument goes (see Kitcher 1993, chapter 8). I aim to re-focus philosophers' attention on the value of *false* theories. Like Feyerabend, I argue that false theories play a crucial role in the development of scientific knowledge. My argument is focussed on a case study from the history of astronomy.

I examine, in detail, the case of astronomy during the period between the late 1580s and the early 1620s. This period begins with the publication of Tycho Brahe's theory, in *On the Most Recent Phenomena of the Aetherial World* (*De Mundi aetherei recentoribus phaenomenis*), and ends after Galileo published his telescopic observations, but before he published the *Dialogue Concerning the Two Chief World Systems: Ptolemaic and Copernican.* I highlight a number of ways in which Brahe's theory changed the debate between early modern Ptolemaic astronomers, and the few early converts to the Copernican Theory. Even though Brahe's theory was not the theory ultimately accepted by the research community, its presence in the debate did enhance scientists' ability to evaluate the competing theories, as Feyerabend suggests. In addition, I draw attention to another important role played by false theories. I argue that Brahe's theory helped facilitate the change of theory, from Ptolemy's theory to Copernicus' theory. Theory change involves two distinct processes: the erosion of a long-held consensus and the emergence of a new consensus. It is only in those cases where there are only two competing theories – the long-accepted theory and the new challenger – that the two processes are indistinguishable. I argue that when there are more than two competing theories to choose from, a radical change of theory can occur in a piecemeal fashion. This facilitates the process of theory change.[1]

4.2   Feyerabend on the Value of Alternative Theories

Feyerabend is most famous for the methodological prescription 'anything goes.' Elisabeth Lloyd argues that this is not Feyerabend's own

---

[1] Larry Laudan defends a piecemeal account of scientific change (see Laudan 1984, chapter 4).

methodological recommendation (see Lloyd 1997, p. S396). Rather, she claims that Feyerabend's point was that for those looking for a binding rule of rationality, 'anything goes' is the only rule that stands the test of time. According to Lloyd, Feyerabend was well aware how empty and useless this rule was (see Feyerabend 1978b, 188; Lloyd 1997, p. S396). Consequently, it is worth distinguishing between this methodological prescription and his endorsement of theoretical pluralism.[2]

More intimately connected with Feyerabend's endorsement of theoretical pluralism is his so-called principle of proliferation. Feyerabend claims that the principle of proliferation is a consequence of his aim to maximise the 'testability of our knowledge' (see Feyerabend 1965/1981, p. 105). The principle of proliferation is a methodological prescription. It is as follows: '*invent, and elaborate theories which are inconsistent with the accepted point of view, even if the latter should happen to be highly confirmed and generally accepted*' (ibid.). Feyerabend presents essentially the same sort of justification in support of theoretical pluralism in *Against Method*. The only difference is that here he talks of hypotheses rather than theories. This difference, though, is irrelevant for our purposes. In *Against Method* he argues that 'there are circumstances when it is admissible to introduce, elaborate and defend ad hoc hypotheses, which contradict well-established and generally accepted experimental results, or hypotheses whose content is smaller than the content of the existing and empirically adequate alternative' (Feyerabend 1975/1988, pp. 14–15). There is much that is packed into this prescription, and it is worth separating the various claims.

First, Feyerabend wants us to see that when choosing to entertain hypotheses or theories scientists *should not* be constrained by the other hypotheses and theories that are widely accepted, even if these hypotheses and theories are 'well-confirmed' (see Feyerabend 1975/1988, p. 20). Such theories, if they are mistaken, would prevent us from working with or accepting a superior theory (or, as a realist might say, one closer to the truth).[3] Theoretical pluralism thus aids scientists in finding better theories

---

[2] Not all commentators on Feyerabend are in agreement here. Jamie Shaw, for example, has recently defended the view that Feyerabend's 'anything goes' is intimately tied with his defence of pluralism (see Shaw 2017).

[3] Feyerabend was a realist. He refers to his position as methodological realism. He believes that when scientists are realists, rather than instrumentalists, they are more apt to get at the truth. Instrumentalism makes scientists complacent, and content with a theory that is merely empirically adequate. On the nature of Feyerabend's realism, see Feyerabend (1964/1981) and K. Brad Wray (2015). Recently, Brown and Kidd (2016) have argued that realism only played a central role in Feyerabend's early work, from 1951 to 1975 (see Brown and Kidd 2016, p. 3). Shaw suggests (in personal communication) that Feyerabend largely eschews talk of truth, focussing instead on

even when the theory they currently accept is empirically successful. Second, Feyerabend wants us to see that when choosing theories, scientists should not be constrained by 'the facts'. The facts, after all, could be contaminated (p. 22). As Feyerabend explains, *'facts are constituted by older ideologies'* (p. 39). Consequently, what is commonly taken to be a fact may not be a fact at all (see Bschir 2015, p. 24). Further, given the effects that the theory one accepts has on what one is apt to observe, as long as a scientist accepts a particular theory some facts may elude detection.

Because of the intimate connection between fact and theory, Feyerabend argues that scientists must often rely on incompatible alternative theories to expose the evidence that might ultimately 'refute a theory' (Feyerabend 1975/1988, p. 20).[4] Indeed, at times Feyerabend makes an even stronger claim: some of the evidence that may refute a long-accepted theory will not be detectable without the aid of another incompatible theory. Underlying Feyerabend's radical view is a conviction that 'some of the most important formal properties of a theory are found by contrast, not by analysis' (Feyerabend 1975/1988, p. 21). By juxtaposing two competing theories, scientists are able to evaluate theories in a more thorough manner.

Feyerabend offers an additional reason in support of theoretical pluralism. He raises the important concern that most methodologies and methodological prescriptions are partial or biased towards the long-accepted theory. They are in this respect epistemically conservative. For example, appeals to what Kuhn calls 'external consistency' – that is, consistency with other accepted theories – are inherently conservative, taking one's current beliefs as a constraint on what new hypotheses one entertains. According to Feyerabend, other things being equal, scientists are inclined to work with the theory they have long accepted, even when faced with an alternative theory that can account for the facts equally as well.[5] But, clearly, were the situation

---

testability. I am inclined to think that when a realist values testability it is, as Popper suggests, because scientists believe testability advances scientists' goals in the pursuit of the truth.

4 The influence that Popper had on Feyerabend can be challenging to detect, given that Feyerabend was often quite critical of Popper and Popperians. His disdain for Popper is expressed most candidly in his correspondence with Imre Lakatos. See, for example, his letter to Lakatos, dated 27 July 1968, reacting to the suggestion that he contributes a paper to a Festschrift for Popper (in Lakatos and Feyerabend 1999, pp. 147–148). Like Popper, though, Feyerabend believed that science progresses when long-held theories are falsified. We never confirm a hypothesis or theory. This is a bit overstated. As Jamie Shaw notes, given Feyerabend's principle of tenacity, 'the very idea of a "cut off point" for degenerating research programs is counterproductive' (see Shaw 2017). Matteo Collodel (2016) provides a comprehensive analysis of Feyerabend's complex relationship to Popper and the Popperians.

5 Kuhn notes that 'as in manufacture so in science – retooling is an extravagance to be reserved for the occasion that demands it' (Kuhn 1962/2012, p. 76).

reversed, if the long-accepted theory was the new theory, our epistemic conservativism would lead us to reject it. Epistemically irrelevant contingencies, such as which theory happened to be developed first, should not determine which theory scientists accept or regard as the better justified theory (see Feyerabend 1965/1981, pp. 119–120).[6]

On the one hand, Feyerabend's rhetoric may be overblown (see Lloyd 1997; Tsou 2003). Developing viable alternative theories is very challenging work. And despite his enthusiasm for theoretical pluralism, Feyerabend is surprisingly silent on how new theories are developed. No doubt, this neglect is partially a consequence of the context of discovery/context of justification distinction, which had the effect of setting matters of discovery and theory generation aside, leaving whatever there is to say about them to sociologists of science (see, for example, Brannigan 1981; see also Hoyningen-Huene 1987). Further, as a matter of historical fact, a research community seldom finds itself in a situation where they are choosing between more than a few viable alternatives, seldom more than three. Here, I have in mind full-blown theories, rather than concepts, comprehensive theories, like the Copernican Theory of the cosmos or Darwin's Theory of Evolution by Natural Selection. But, on the other hand, theoretical pluralism does appear to deliver the benefits that Feyerabend identifies, even when there are only three or four competing theories to choose from. In Feyerabend's words, *'variety of opinion is necessary for objective knowledge'* (Feyerabend 1975/1988, p. 32).

4.3 Feyerabend on Brahe

Feyerabend often appeals to the Copernican Revolution in astronomy to illustrate his more general points about methodology. For example, he mentions how Copernicus' theory was contrary to many accepted 'facts', as well as 'auxiliary sciences' (see Feyerabend 1975/1988, p. 52). That is, Copernicus' astronomical theory was contrary to the accepted Aristotelian terrestrial physics. The tower argument allegedly demonstrated this conflict. Were the Copernican theory correct, an object dropped from a tower would land far from the base of the tower, given

---

[6] W. V. Quine has defended the sort of epistemic conservativism that Feyerabend attacks. Quine recommends the methodological maxim of minimal mutilation: faced with new evidence one should minimise the mutilation to one's current belief system (see Quine 1992). Jonathan Tsou notes that Feyerabend's attack on the consistency condition, which 'makes a well-accepted theory *the* measure for the introduction of a new theory', is based on a concern about the pernicious effects of epistemic conservativism (see Tsou 2003, p. 212).

the base of the tower would have moved some distance in virtue of the Earth's daily rotation on its axis in the time the object took to reach the ground.[7] Feyerabend also discusses Galileo's innovations and their misfit with the accepted theories, at some length. For example, Feyerabend notes that Galileo '*introduces a new observation language*' in his efforts to defend the Copernican theory, and to break out of the 'common idioms . . . of the Aristotelian philosophy' and its understanding of motion (see Feyerabend 1975/1988, pp. 64–65).

Surprisingly, though, Feyerabend seems to pass over the period between Copernicus and Galileo, thus neglecting to discuss Brahe's contributions. In many ways, this is rather unfortunate as Brahe provides a clear case of an alternative theory, in this case a third well-developed theory for astronomers to consider. Consequently, Feyerabend could have put to good effect an analysis of Brahe's theory and its impact on the history of early modern astronomy in his effort to defend theoretical pluralism.

Feyerabend, though, does make a few passing remarks about Brahe's contributions to astronomy. For example, Feyerabend describes Brahe as one of a number of 'outstanding astronomers' who did not accept Copernicus' theory (see Feyerabend 1981e, p. 13). Feyerabend's point is that, contrary to the standard narrative, Copernicus' theory was not so compelling as to immediately bring all right-minded astronomers around to accepting it. The astronomers who resisted Copernicus' radical innovations were not all dogmatic Aristotelian university professors and Jesuits. Feyerabend also mentions Brahe as one who objected to the equant point in Ptolemaic planetary models, and regarded it as a rationale for 'rebuilding astronomy' (see Feyerabend 1981d, p. 243). The equant point shifted the centre of motion of a planet away from the centre of the circle around which the planet moved, and was thus thought by some to constitute a violation of a key principle in astronomy, that all motion is circular or composed of a combination of circles. Ptolemy's employment of the equant was a pressing concern for Copernicus as well (see Copernicus 1543/1995, pp. 12, 283; see also Feyerabend 1975/1988, p. 144).[8] Further,

---

[7] Copernicus discusses the motion of falling bodies, and Galileo addresses the tower argument, attempting to show that it does not have the implications that Ptolemaic astronomers suggest for the heliocentric theory (see Copernicus 1543/1995, Book 1, chapter 8; Galilei 1632/2001, pp. 161–164). Copernicus claims that 'the movement of falling and rising bodies is twofold and is in general compounded of the rectilinear and the circular' (Copernicus 1543/1995, p. 17). Apparently, Brahe found the tower argument persuasive (see Schofield 1981, pp. 81–82).

[8] Copernicus explicitly discusses Ptolemy's use of the equant in presenting his model of the orbit of Mercury (see Copernicus 1543/1995, p. 283). None of Copernicus' planetary models employ equant points. In chapter 4 of Book One of *On the Revolutions*, Copernicus makes clear his commitment to

Feyerabend mentions Brahe as a contributor to the Copernican Revolution in astronomy, even though he did not accept *Copernicus' theory* (see Feyerabend 1975/1988, p. 139). How exactly Brahe contributed to the revolution, Feyerabend does not specify. He does note that Brahe's '*observations* contributed to the downfall of generally accepted views' (p. 139; emphasis added). But, it seems much more could be said about Brahe's contributions to the Copernican Revolution, especially as part of Feyerabend's defence of *theoretical* pluralism.

4.4 The Significance and Effects of Brahe's Theory

I want now to consider Brahe's contributions to the Copernican Revolution in a more detailed and systematic way. I aim to draw attention to another way in which theoretical pluralism aids scientists in the pursuit of their epistemic goals, one that focusses on the value of false theories.

Brahe's contributions to early modern astronomy are quite extensive. He is rightly famous for his advances in methodology in observational astronomy that profoundly changed the practice of astronomy in a number of important ways. He employed larger and more accurate instruments than his predecessors employed (see Brahe 1598/1946, pp. 65, 79; Gade 1947, pp. 84–85; Thoren 1990, pp. 190–191). The size of some of the instruments required a team of workers. Brahe describes how certain of his instruments required two observers to operate them, including the Bipartite Arc and the Triangular Astronomical Sextant, (see Brahe 1598/1946, pp. 70, 74). And some were even more complex, requiring 'a seasoned observer, a team leader to read off the positions of sightings, a secretary to record observations, and sometimes a clockwatcher' (see Christianson 2000, p. 80). Brahe could afford to build these instruments and retain the required personnel because of his wealth and the generous support he received from the Danish crown (see Brahe 1598/1946, p. 109; Heilbron 1992, p. 45; Christianson 2000). He collected far more observations than his predecessors collected (see Brahe 1598/1946, pp. 112–113). At Uraniborg, Brahe's island estate, they conducted 'about 85 observing sessions a year' (Thoren 1990, pp. 201, 220).

circular motions in modelling the motions of the planets. 'We must ... confess that [the planets'] movements are circular or are composed of many circular movements, in that they maintain [their observed] regularities in accordance with a constant law and with fixed periodic returns; and that could not take place, if they were not circular. For it is only the circle which can bring back what is past and over with' (p. 12).

And Brahe instituted practices that were explicitly designed to detect errors in and ensure the reliability of his observations. For example, he would have two teams of astronomers collecting observations of the same star at the same time at different locations on his estate, thus providing a check on each team's observations (see Gade 1947, pp. 69, 90). Brahe took other measures to ensure the accuracy of his instruments, including '[assessing] their performance relative to one another' (Mosley 2007, p. 60). According to Allan Chapman 'crucial constants, such as the solar altitude or First Point of Aries, would be observed with a battery of instruments, both to obtain the best average value, and to cross-check the strengths and weaknesses of different designs' (Chapman 1989, p. 74). Elaborating, Chapman notes that Tycho's 'constant revision of the same fundamental observations enabled him to get average values that vastly exceeded the angular resolution of the naked eye' (p. 75). That is, by taking many observations with a variety of different instruments and averaging these, Brahe apparently converged on values that exceeded the accuracy of his instruments.

These innovations played an important role in developments in early modern astronomy. The data collected by Brahe played an indispensable role in the development of Kepler's first two laws of planetary motion (see Kepler 1609/2004, p. 8 fn. 1; Christianson 2000, pp. 304–305; Mosley 2007, p. 27). This is widely recognised.

But, Brahe also developed his own theory, different from the theories of Ptolemy and Copernicus. This is worth considering in light of Feyerabend's defence of theoretical pluralism. Brahe's theory is often described as a 'geoheliocentric' theory. As in Ptolemy's theory, in Brahe's theory the Moon, the Sun and the Stars orbit the Earth each day. But in Brahe's theory the various planets – Mercury, Venus, Mars, Jupiter and Saturn – orbit the Sun as they are swept around the Earth each day with the Sun. In this respect, Brahe's theory was more like Copernicus' theory than Ptolemy's theory. The Sun was the centre of motion for the planets.

It is worth considering how the introduction of this new theory affected the debate in astronomy. In the remainder of this section, I will argue that theoretical pluralism plays a crucial role in two dimensions of theory change that are often not clearly distinguished: (i) the erosion of a consensus on a long-accepted theory, and (ii) the shift of allegiance in a research community to a new theory.

Strictly speaking, there was not much of a debate about cosmology in the 1580s when Brahe developed and published his new theory. There was a firm consensus around the Ptolemaic theory. Few astronomers regarded the

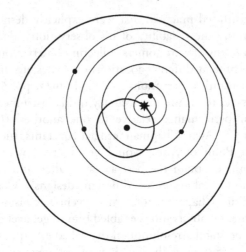

Figure 4.1 The Tychonic system: the Earth is at the centre.

Copernican theory as a true description of the cosmos. In fact, by 1600 only ten astronomers were realists about the Copernican cosmology. According to Robert Westman, 'four were German (Rheticus, Michael Maestlin, Christopher Rothmann and Johannes Kepler); the Italians and English contributed two each (Galileo and Giordano Bruno; Thomas Digges and Thomas Harriot); and the Spaniards and Dutch but one each (Diego de Zuñiga; Simon Stevin)' (Westman 1986/2003, p. 54). But, as Westman notes, a number of other astronomers *used* the Copernican theory, or at least the planetary models. These are the so-called Wittenberg School of Astronomers (see Westman 1975). But, the Ptolemaic theory was still unquestionably the dominant theory. So, Brahe did not introduce his new theory because European astronomers were frantically looking for an alternative theory. Indeed, given the firm grip that Aristotelian physics had on European intellectuals, and the fact that Ptolemy's theory fit so well with it, it is not surprising that astronomers were reticent to adopt Copernicus' new cosmology.

But, one important effect that the publication of Brahe's theory had was to make the choice astronomers faced less polarised. It did this in a number of ways. First, the choice was no longer between the Ptolemaic theory and the Copernican theory. And with each new theory of the cosmos, the under-determination of theory choice became more evident. Astronomers had long known that various mathematical devices – equant points and epicycles and deferent circles – could be used to model the same effects (see, for example,

Copernicus 1543/1995, p. 5; see Shank 2002, pp. 183–184). But, these new *theories* made it clear that the question of the correct *cosmology* was underdetermined by the evidence as well. This awareness would have probably contributed to the erosion of the conviction that Ptolemy's theory *must* be correct. The logical possibility that there is an alternative theory that could account for the relevant data rarely moves scientists. In fact, even Kyle Stanford's (2006) concern, that scientists are (almost) always in a state of underdetermination because of the existence of unconceived alternatives that will, in the future, prove to be superior to the accepted theory, is unlikely to erode the working scientists' confidence in the accepted theory, as long as it is empirically successful. But, when genuine competitors are developed, scientists cannot help but consider the *possibility* that they have settled on the wrong theory. Indeed, Ernst Mach believed that instruction in the history of science could play a similar role. When scientists learn about the various theories accepted in the past, they are apt to be less dogmatic, which is good for science (see Mach 1911, p. 17).

Second, Brahe's theory showed in a very concrete way that specific hypotheses about the structure of the cosmos were independent of each other, at least in principle. Astronomers could see that the choice was not between two wholly incompatible theories: A and B. To appreciate this point, consider the following four hypotheses:

1. The Earth is the centre of the cosmos,
2. The motion of the planets are tied physically to the Sun,
3. Mars is sometimes closer to the Earth than it is to the Sun and
4. The Earth rotates on its axis in 24 hours.

Ptolemy accepted the first hypothesis, but not the second, third, or fourth hypotheses.[9] Copernicus accepted the second, third and fourth hypotheses, but not the first. And Brahe accepted the first three hypotheses, but not the fourth. I am not claiming that seventeenth-century astronomers could not conceive of these hypotheses separate from the particular clusters in which they were gathered together in the Copernican and Ptolemaic theories until Brahe published his theory. Rather, my point is that Brahe showed his scientific peers in a very concrete way how different hypotheses could be

---

[9] Through a variety of contrivances in Ptolemy's planetary models, the planets' motions were tied to the Sun. For the inferior planets, Mercury and Venus, Ptolemy stipulated that the centre of the epicycles of the planets were always on the line running from the Earth to the Sun. For the superior planets, Mars, Jupiter and Saturn, Ptolemy stipulated that the line running from the planet to the deferent circle was always parallel to the line running from the Sun to the Earth. These constraints were built into Ptolemy's planetary models ad hoc, to reconcile the predictions from his theory with observations, some of which he inherited from his predecessors.

Table 4.1 *Five competing astronomical theories and their relationship to four hypotheses*

	Earth-centred	Planets orbit the Sun	Mars closer to the Earth than the Sun	The Earth rotates on axis
Ptolemaic theory	yes	no	no	no
Copernican theory	no	yes	yes	yes
Tychonic theory	yes	yes	yes	no
Ursus' theory	yes	yes	no	yes
Egyptian theory	yes	no	no	no

* In the Egyptian theory, only the inferior planets, Mercury and Venus, orbit the Sun. The superior planets, Mars, Jupiter and Saturn, orbit the Earth.

combined together in an empirically successful theory that was neither Copernican nor Ptolemaic. That is, Brahe helped astronomers see that these hypotheses are genuinely separable, and logically distinct.

Let us consider this third hypothesis in more detail. Apparently, 'the distance of Mars became "a central driving theme" of Tycho's observational programme, beginning in 1582' (Gingerich and Voelkel 1998; Christianson 2000, p. 73). Initially, what motivated Brahe was the realisation that the third hypothesis – the hypothesis about the distance between Mars and the Earth – provided an ideal test between the Ptolemaic theory and the Copernican theory. Brahe realised that

> the two hypotheses [that is, the Ptolemaic and Copernican theories] are in sharp contradiction over the relative distances from the Earth of the Sun and the planet Mars at the times when Mars is in opposition to the Sun . . ., i.e. when Mars, the Sun and the Earth are all situated in the same straight line, the Earth being between Mars and the Sun. According to the Ptolemaic hypothesis Mars must at these times be further from the Earth than is the Sun, while according to the Copernican hypothesis it must be nearer to the Earth than is the Sun. (Schofield 1981, pp. 56–57)

This provided a crucial test of sorts for the two theories, in a way that Galileo regarded the phases of Venus as a crucial test after the invention of the telescope.[10] Specifically, Brahe realised that he may be able to determine

---

[10] As Pierre Duhem notes, there really are no crucial tests in science (see Duhem 1914/1954, chapter VI). But a failed prediction on the part of a theory really does shift the burden of proof in disputes between advocates of competing theories.

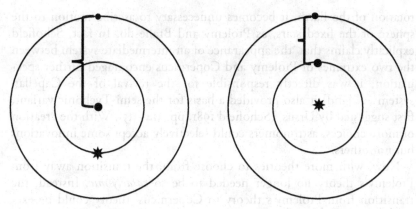

Figure 4.2 The Ptolemaic and Copernican systems: the arrangement of Mars, the Sun and the Earth when Mars (at the top) is in opposition to the Sun. Note: Mars is at the top of the diagram in both the diagrams.

which theory is correct by measuring the diurnal parallax of Mars, 'the apparent difference in position created by viewing an object at its zenith versus the position observed with the object on the horizon' (see Gingerich and Voelkel 1998, p. 3). In time, Brahe realised that his own theory also implied that Mars is sometimes closer to the Earth than it is to the Sun. Indeed, there is reason to believe that Brahe developed his new theory, in part, because of his belief that Mars is sometimes closer to the Earth than it is to the Sun.

Third, as Christine Jones Schofield argues, 'Tycho's ideas ... encouraged freer speculation on the system of the world, and the introduction of new hypotheses rather than reliance upon ancient ones' (Schofield 1981, p. 144). That is, in proposing his new theory of the structure of the cosmos, Brahe showed that the space of alternatives was larger than previously thought. This may have contributed to the development of other theories as well. In fact, astronomers were faced with even more alternatives by the early seventeenth century. The so-called Egyptian model, attributed to Martianus Capella, was revived and defended by some (e.g. Schofield 1981, p. 175). This theory is similar in many respects to the Ptolemaic model, except that Venus and Mercury orbit the Sun, rather than the Earth (see Schofield 1981, p. 172). And Brahe's rival, Nicolaus Reymers Bär, developed a system that was essentially the same as Brahe's system except that the Earth turned on its axis daily (see Mosley 2007, p. 177).[11] With the daily

---

[11] Brahe suggested that three astronomers 'have not been ashamed to appropriate [his] hypothesis and present it as their own invention' (Brahe 1598/1946, p. 115). Raeder, Strömgren and Strömgren identify them as Reymers, Liddel and Röslin (ibid).

rotation of the Earth, it becomes unnecessary to ascribe motion to the sphere of the fixed stars, as Ptolemy and Brahe do. In fact, Schofield explicitly claims that 'the appearance of an intermediate system between the two extremes of Ptolemy and Copernicus encouraged further speculation; it was directly responsible for the revival of the Capellan system ... [and] it also provided a basis for the semi-Tychonic variant, first suggested by Ursus' (Schofield 1981, pp. 314–315). With the creation of more choices, astronomers could selectively accept some innovations but not others.

Now, with more theories to choose from, the transition away from Ptolemy's theory no longer needed to be so *cataclysmic*. Instead, the transition from Ptolemy's theory to Copernicus' theory could be executed in a more piecemeal way, via Brahe's theory, or some other alternative. The flexibility thus afforded by these alternative theories probably contributed significantly to the revolution in early modern astronomy. The *erosion of the consensus* no longer required the adoption of a single specific competitor theory. Rather, different astronomers could be compelled to abandon the Ptolemaic theory by different evidential considerations and thus be moved to accept or work with different theories. Theory change involves both the abandonment of a long-accepted theory, and the acceptance of a new theory. When there are only two competing theories to choose from, these distinct processes can seem like one and the same process. In such cases, evidence must both erode one's commitment to one theory, and compel one to accept another theory. When there are only two theories to choose from, the cataclysmic nature of the change required may discourage some from adopting the new theory.

Fourth, Brahe's *theory* also contributed to developments in instrumentation and observational practices, as he sought to find proof for his own theory, or at least some sort of proof against the Ptolemaic Theory. This is vividly illustrated in Gingerich's and Voelkel's account of Brahe's attempt to detect a predicted diurnal parallax of Mars. The measurements required were very subtle and required a degree of precision that tested the limits of even Brahe's instruments (see Gingerich and Voelkel 1998). Thus, the development of alternative theories fed back into the development of other aspects of astronomy that in turn contributed to the generation or collection of new data. Of course, Brahe had other motives for improving his instruments, but the particular research projects that occupied him would, no doubt, affect which refinements were given priority. And his interest in detecting a Martian parallax focussed his attention on some instruments rather than others (see Gingerich and Voelkel 1998, p. 13).[12]

In closing, it is worth noting that Brahe held a view similar to Feyerabend's about the value of *comparing* competing hypotheses and theories. In the Preface to his published collection of letters, *Epistolae astronomicae*, Brahe notes that 'the different opinions of others about the same things are more correctly recognized and evaluated when someone frankly and freely shares with a friend in letters that which he has ascertained and established on a topic he has studied . . . from this friendly and frank collation, truth is that much more readily and clearly extracted' (Brahe 1596, in Mosley 2007, p. 138).[13] Though Brahe was often himself quite secretive about his research, his extensive correspondence with the leading astronomers of his day certainly attests to the fact that he believed that there was a great benefit to 'collating different opinions' on a research topic.

4.5 Theoretical Pluralism Today

Feyerabend's provocative rhetoric did much to alienate him and his views in mainstream philosophy of science. He is frequently cast as an enemy of rationality (see Stove 1982; Theocharis and Psimopoulos 1987; Haack 2007). But, despite this, some philosophers of science now take theoretical pluralism quite seriously. Still it seems that much of the contemporary interest in theoretical pluralism is largely motivated by the concern to ensure that there is an effective division of labour in the scientific research community, with all *viable* hypotheses being pursued. This is most evident in the work of Philip Kitcher who is explicitly concerned with shedding light on how a community of scientists can collectively realise their epistemic goals even if they do not explicitly or deliberately coordinate their efforts to do so (see Kitcher 1993, chapter 8). Only those hypotheses and theories that might in fact be true are deemed to have value. Only they warrant the consideration of scientists. The value of pluralism on this view

---

[12] Chapman provides a clear account of the various instruments Tycho employed, distinguishing between three types, each serving a different purpose: (i) Sextants 'were used for measuring either vertical or horizontal angles between pairs of objects in the sky' (Chapman 1989, p. 71); (ii) Armillaries enabled one to 'observe Right Ascension angles, measure the Sun's daily coordinates and fix the First Point of Aries' (p. 72); and (iii) Quadrants enabled one to 'measure vertical angles between horizon and zenith' (p. 73).

[13] Mosley uses this quotation to support his claim that Brahe believed that *letters* are a valuable resource for understanding disputes in astronomy, and in particular for determining the virtues and weaknesses of competing theories.

is that it ensures that the true theory is not lost, as most scientists spend their time working with a theory that is empirically adequate, but false. This is not quite Feyerabend's concern, for he is explicit that even false theories, outrageous theories, even, can be valuable.

Even some of Feyerabend's sympathetic commentators tend to emphasise the fact that hitherto unconsidered alternatives may be true, as if it is the likelihood of an alternative theory being true that constitutes the value of theoretical pluralism for Feyerabend. Jonathon Tsou, for example, claims that Feyerabend appeals to two principles to support his theoretical pluralism.

> 'The Anti-Uniformity Principle' ... [states that] Principles that encourage uniformity in science have a dual effect of protecting currently accepted theories and hindering the discovery of new theories that are potentially better than currently accepted theories. (Tsou 2003, p. 214)

And

> 'the Proliferation Principle' (PP) ... [states that] Principles that encourage a plurality and proliferation of scientific theories have the effect of criticizing already accepted theories and encouraging the discovery of new theories that are potentially better than already accepted theories. (p. 216)[14]

Again, this is only one part of Feyerabend's defence of pluralism.

Hasok Chang has recently offered a sustained defence of pluralism and his pluralism is more in line with Feyerabend's (see Chang 2012, chapter 5). Consequently, it is worth examining Chang's defence in some detail, and comparing it with Feyerabend's defence of theoretical pluralism. This will provide me with an opportunity to bring into focus exactly what is unique or especially insightful about Feyerabend's theoretical pluralism.

Chang's project is broader than Feyerabend's. Chang is not concerned exclusively with *theoretical* pluralism, which is not surprising, given his leadership role in the philosophy of scientific practice movement. Instead, Chang recommends a pluralism with respect to what he calls 'systems of practice' (see Chang 2012, p. 260). A system of practice is 'a coherent and intersecting set of epistemic activities performed with a view to achieve

---

[14] Oddly, Tsou's characterisation of Feyerabend's Principle of Proliferation differs from Feyerabend's. According to Feyerabend, the *principle of proliferation* makes the following recommendation: '*Invent, and elaborate theories which are inconsistent with the accepted point of view, even if the latter should happen to be highly confirmed and generally accepted*' (Feyerabend 1965/1981, p. 105; emphasis in original). Elaborating, Feyerabend notes that 'the principle of proliferation not only recommends invention of new alternatives, it also prevents the elimination of older theories which have been refuted. The reason is that such theories contribute to the content of their victorious rivals' (p. 107).

certain aims' (ibid). Of course many, maybe all, scientific systems of practice will include or implicate a theory, but a system of practice is something broader than a theory. But, it is worth noting that the case studies that figure in Chang's defence of pluralism involve competing theories, for example, the phlogiston theory and Lavoisier's competing theory, or competing ways of operationalising theoretical entities, for example, the chemical atom (chapter 1 and section. 3.21). So theoretical pluralism is central to Chang's concerns.

Feyerabend is also not exclusively concerned with theoretical pluralism. Rather, his defence of theoretical pluralism was part of a broader goal, a defence of methodological pluralism. But my focus here has been narrowly with assessing theoretical pluralism, and I want to stay focussed on theoretical pluralism.

Chang offers two different types of defences of pluralism: (i) appeals to the 'benefits of tolerance' and (ii) appeals to 'benefits of interaction' (see sections 5.2.2 and 5.2.3). It seems that some of the considerations he raises regarding the benefits of pluralism are explicitly concerned with not losing the truth. For example, he notes that pluralism lets scientists hedge their bets (see section 5.2.2.1). In this way, the truth is less likely to be lost. In this respect, Chang's position is closer to Kitcher's position than to Feyerabend's (see section 5.2.2).

Further, Chang's defence of pluralism also appeals to pragmatic considerations. For example, he argues that one reason we should not discard a theory once we discover its limitations, as the monist seems to recommend, is that 'workable systems are not easy to come by, and they should be preserved as much as possible' (p. 258). So even though we may have developed a superior theory, the workable system may still serve some goals (see p. 260). None of this is objectionable, but such pragmatic concerns are tangential to the epistemic concerns that motivate Feyerabend in his defence of theoretical pluralism. That is, Feyerabend is concerned exclusively with the epistemic merits of pluralism.

As we saw earlier, Feyerabend has a different motivation for his defence of theoretical pluralism. Even theories that are very unlikely to be true or even close to the truth can play a valuable epistemic role. In fact, he claims that *'there is no idea, however ancient and absurd, that is not capable of improving our knowledge'* (Feyerabend 1975/1988, p. 33). Feyerabend's discussions of voodoo and astrology are clear indicators that he wanted to draw attention to the value of even clearly false theories (on voodoo, see pp. 35–36). To be clear, in the 1590s, it was not at all obvious that Brahe's theory was a false theory. It is only with hindsight that we can make that judgement. It was by no means on a par with our present day assessment of voodoo (about which I know very little). Alternative theories, though, ensure that competing theories are more thoroughly evaluated than they would be otherwise. Because of the comparative nature of theory evaluation, competing theories aid us in attending to aspects of a theory that we may otherwise overlook. This is the Millian dimension of Feyerabend's project, emphasised by Lloyd (Lloyd 1997, §§ 3–4; also Feyerabend 1975/ 1988, pp. 30–31; 1981c, pp. 139–143). In *On Liberty*, John Stuart Mill argued that false views can assist in keeping the truth alive, by ensuring that people understand why they hold the views they hold (see Mill 1859/1956, p. 21).

My aim was to develop the defence of theoretical pluralism even further. I have argued that competing theories also play a role in providing a means for a community of scientists to change their allegiance from one theory to another in a piecemeal manner. As a result, evidence can function in a subtle way, eroding the prevailing consensus, and creating a new consensus, often in a multi-step process, in which some scientists pass through an alternative theory that never does hold the allegiance of the research community as a whole. Traditionally, theory change has been discussed in quite abstract terms, as merely involving a decision to select one of two or more options. The only relevant consideration is the evidence, and perhaps the theoretical virtues, if the evidence does not speak clearly enough in favour of one of the choices. My analysis shifts the focus. When scientists are faced with the choice between competing theories, they are often making a choice from a particular theoretical point of view, one that they have long taken for granted, due to their early training, and years of working successfully with that theory. Such real-life scientists may resist a radical change of theory. Many may need to make the shift to a new radical alternative in a piecemeal manner, first accepting a less radical alterative, before ultimately accepting the radical new alternative. Theoretical pluralism makes such transitions possible.[15]

[15] Acknowledgements: I thank Jamie Shaw, Lori Nash and Line Edslev Andersen for critical feedback on earlier drafts.

CHAPTER 5

Epistemological Anarchism Meets Epistemic Voluntarism
Feyerabend's Against Method *and van Fraassen's* The Empirical Stance

Martin Kusch

5.1 Introduction

In this chapter, I shall compare and contrast central themes of Paul Feyerabend's best-known work, *Against Method* (1975a, subsequently AM) with pivotal ideas of Bas van Fraassen's 2002 book, *The Empirical Stance* (subsequently ES). The comparison appears fruitful for two reasons: first, because van Fraassen is one of the few contemporary philosophers of science who continue to engage closely and charitably with Feyerabend's work; and, second, because van Fraassen disagrees with some of Feyerabend's central contentions. Here, I do not have the space to determine conclusively who of the two philosophers is right where they take different views on a given question; I shall be satisfied to clearly identify the issues and disputes that need further reflection.

The scope of this investigation is restricted primarily to ES and the first edition of AM. I look beyond these two books only where ES's discussion of Feyerabend's views draws on texts other than AM, and where van Fraassen further develops important claims of ES. I have elsewhere discussed Feyerabend's relativism in his later writings (Kusch 2016). Moreover, Feyerabend's and van Fraassen's respective oeuvres are of such breadth, depth and development over time, that a full consideration of all their important similarities and differences would require a book-length treatment.

Two ideas from ES will be crucial in what follows. The first idea is that many philosophical positions are best rendered not as 'doctrines' but as 'stances', that is, as sets, systems or bundles of values, emotions, policies, preferences and beliefs. (To avoid torturous repetition, I shall refer to sets of values, emotions, policies and preferences as 'VEPPs'.) The second idea is a form of epistemology van Fraassen calls 'epistemic voluntarism'. It is based on the rejection of two received views: that the principles of rationality determine which philosophical positions and scientific paradigms we must adopt, and that epistemology is (akin to) a descriptive-explanatory (scientific) theory of cognition.

I shall first argue that Feyerabend's 'epistemological anarchism' is best rendered as a stance rather than a dogma. Subsequently, I shall explore similarities and differences between van Fraassen's epistemological voluntarism and Feyerabend's epistemological anarchism. I shall propose that Feyerabend's position is more radical than van Fraassen's position. In the process, I shall also discuss the two philosophers' views on scientific revolutions, rationality and relativism. My main interest in all this is to arrive at a better understanding of anarchism and epistemic relativism in the philosophy of science.

5.2   Stances and Dogmas in ES

Empirical Stance suggests that many philosophical positions are best understood as 'stances', that is, as bundles of VEPPs, rather than as beliefs, theories or 'dogmas' about the world. There are three main motivations behind this proposal. The first motivation is specific to empiricism. If we seek to capture the 'core idea' of empiricism in the dogma (D) 'experience is the one and only source of information' (ES, p. 43), then it is difficult to explain how empiricists can reject their main rival, that is, metaphysics.

The difficulty is the result of three commitments 'dogma-empiricists' (=empiricists$_D$) are ready to make. First, the dogma is meant to licence an outright rejection of metaphysics, without any further, detailed argumentative give-or-take over details of specific meta physical proposals. Second, empiricists$_D$ hold that all admissible philosophical theses must be empirical hypotheses. And third, like all empiricists, so also empiricists$_D$ admire the ideal of empirical science according to which all empirical hypotheses must be rigourously tested by observation and experimentation. Unfortunately, these commitments do not cohere. If D is an empirical hypothesis, then so are statements contradicting D. Some of the latter are core doctrines of alternative philosophical positions, like metaphysics. But, if these rival doctrines too are empirical hypotheses, then they too deserve to be tested in the normal way of empirical science. In other words, these alternative doctrines cannot be ruled out on the sole the basis that they conflict with D.

For ES all this is reason enough to give up empiricism$_D$. Van Fraassen's empiricism (=empiricism$_{vF}$) is defined not by a single dogma but by a set of VEPPs and beliefs. This gives the admiration for the ideal of empirical-scientific inquiry a new role in the opposition to metaphysics. Empiricism$_{vF}$ can rule out metaphysics immediately, and without entering extended debate. This is because metaphysics does not emulate the conduct of empirical science; it does not value empirical hypotheses; and it does not engage in the empirical testing of philosophical theses. We have here an incompatibility in VEPPs rather than an incompatibility in dogmas. But, it still gives the empiricist$_{vF}$ a sufficient and immediate reason for rejection or rebellion.

Making VEPPs central to our understanding of empiricism naturally leads to the idea that many philosophical positions are 'stances' rather than 'dogmas'. What characterises empiricism$_{vF}$ is a

> rejection of explanation demands and dissatisfaction with and disvaluing of explanation by postulate ... calling us back to experience ... rebellion against theory, ideals of epistemic rationality ... admiration for science, and the virtue they see in an idea of rationality that does not bar disagreement ... The attitudes that appear in these list are to some extent epistemic and to some extent evaluative, and they may well involve or require certain beliefs for their own coherence. But none are equatable with beliefs. (ES, p. 47)

Van Fraassen's other two motivations for moving from dogmas to stances are not worked out in similar detail and remain largely implicit in ES. The second can be put as follows. To attribute an 'ism' – like empiricism or materialism–*as a stance rather than a dogma* is to make a claim about where one should locate the *core* or *essence* of the respective position. Do the advocates of the position foreground a claim or doctrine (about the world), and do they marshal their evidence and VEPPs as arguments for this doctrine? Or do the advocates put the emphasis on rebellion, admiration and VEPPs, and treat stated credos as no more than rough glosses intended to point beyond themselves at the underlying epistemic and evaluative commitments?

The third motivation is van Fraassen's desire to find a common denominator for the numerous self-proclaimed empiricists in our tradition, from the Ancient Greek school of physicians, the '*Empirici*', to the Vienna Circle and beyond. A short dogma like D does not suffice for this task. Invoking the rebellion against metaphysics (and related VEPPs) is much better to identify a crucial commonality.

5.3   Epistemological Anarchism as a Stance

Most interpreters and critics of AM have tried to sum up epistemological anarchism in a snappy dogma: '*anything goes*' (AM, p. 28). This dogma is taken to be the central and only claim the epistemological anarchist makes about scientific work. With a 339-page book thus reduced to two words, it is then easy to demonstrate that AM is obviously not to be taken seriously (cf. Shaw 2017). After all, 'anything goes' is absurd both as a descriptive and as a prescriptive statement. Clearly, in science as in life, *everything does not go* at every stage: some courses of action are downright silly or immoral; they involve the actor or deliberator in self-sabotage; or they are physically impossible to carry out. Given the obvious truth of this descriptive statement, 'anything goes' cannot be a meaningful prescription either. Moreover, it is conceptually impossible to follow a rule that makes no distinctions. We are not dealing with a rule unless we are told what is to be done, or not to be done, under specific conditions.

It is striking how often 'anything goes' is still used, by friends and foes of epistemic relativism alike, in this entirely negative way. *Anti-relativists* employ the formula to identify what they see relativism reducing to; *relativists* themselves find it useful as an *absurd and debilitating* foil from which to distinguish their allegedly more *benign* versions of the position.

Interpreting epistemological anarchism as a stance rather than a dogma is one – though perhaps not the only one – way of rendering AM with at least a modicum of interpretative charity. Taking our lead from van Fraassen's discussion of empiricism, the natural starting point is to recognise that anarchism too is primarily a *rebellion*. But anarchism rebels not against metaphysics but against all (primarily philosophical) measures seeking to restrict human freedom. In this vein, Feyerabend attacks attempts to regiment scientific work by tying it to a specific 'logic', by restraining 'imagination', by standardising language and by denying the input from 'religion', 'metaphysics' or 'sense of humour' (AM, p. 19). In philosophy, the anarchist refuses to accept established philosophical dichotomies, for instance, reason versus anti-reason, or sense versus non-sense (p. 191).

Against Method also speaks of 'the stultifying effect of "the Laws of Reason"' (p. 20) or the attempts of "critical rationalism" to turn the scientist into "a miserable, unfriendly, self-righteous mechanism without charm and humour"' (p. 175). The 'one thing' the anarchist always 'opposes positively and absolutely' is universal standards, universal laws, universal ideas like 'Truth,' 'Reason,' 'Justice,' 'Love' and the behaviour they bring along, though he does not deny that it is often good policy to act as if such laws (such standards, such ideas) existed, and as if he believed in them. (p. 189)

Turning from rebellion to commitments, epistemological anarchism is defined by the VEPPs of 'humanitarianism' (p. 19), Millian 'liberalism' (p. 23) and the ideal of the 'mature citizen' (p. 309). The anarchist commits to 'not hurt[ing] a fly—let alone a human being', to 'taking things lightly' and to engaging in 'joyful experiments' (p. 23). Above all else, she values the right and duty to go against established rules and restrictions whenever this strikes her as hindering epistemic or social progress. 'Anything goes' is a rough gloss on these VEPPs. In particular, it expresses the policy of not accepting any a priori or in-principle restrictions on what is the right course of action in novel circumstances. This is underwritten by observations on the history of science. Even restricting ourselves to what we regard today as good science in the past and present, we find that this science is not only the product of reason, but also often of unreason; not only of proof but also often of propaganda; not only of methodological standards, but also of throwing such standards overboard.

Since epistemological anarchism does not have a 'programme' beyond the just-mentioned commitments, its advocates are free to pick opposite sides on one and the same (scientific) issue. Feyerabend regards this idea of freedom as close kin to 'Dadaism', quoting an art-historian's summary that 'Dada ... not only had no programme, [but] ... was against all programmes' and that 'to be a true Dadaist, one must also be anti-Dadaist' (p. 189). The anarchist has 'no everlasting loyalty to, and no everlasting aversion against, any institution or any ideology' (p. 189). Thus the seventeenth-century epistemological anarchist might defend either Bellarmine or Galileo, depending on whether she gives greater weight to the religious needs of 'the common man' or to the opposition to 'orthodox ideology' (p. 193).

The fact that epistemological anarchists can, and typically will, choose opposite sides of the same scientific issue is important for the existence of scientific 'pluralism:' the coexistence of scientific theories that contradict one another. Indeed, AM is adamant that pluralism is ultimately underwritten by humanitarianism, the system of VEPPs at the heart of epistemological anarchism. Here Feyerabend repeatedly emphasises his common cause with John Stuart Mill's *On Liberty* (pp. 43, 53).

To sum up: AM's epistemological anarchism is better taken as a stance (of VEPPs) than as a dogma (like 'anything goes'). In this respect at least, it pays to read AM through the perspectives offered by ES. *Against Method* is strengthened when interpreted as a stance; and the ES's theorising about philosophical positions is confirmed, at least to some extent, by the fact that a philosophical position other than empiricism can fruitfully be rendered as a stance. At the same time, the proximity between ES and AM on this point also raises an important follow-up question: how does epistemological anarchism relate to van Fraassen's epistemic voluntarism? I next turn to addressing this question.

5.4   Epistemic Voluntarism and Scientific Revolutions in ES

Epistemic voluntarism is based on two central claims:

1. Principles of rationality underdetermine our choice of philosophical stances or scientific paradigms and
2. A theory of epistemic rationality must not be 'objectifying,' that is, it must not be a descriptive–explanatory theory of cognition.

By 'principles of rationality' van Fraassen means first and foremost deductive logic, the theory of probability and the practical syllogism. As long as we stick to these principles, we avoid inconsistency and incoherence; we avoid reasoning in ways that – even by our own lights – results in 'self-sabotage': a reasoning that prevents us from reaching our goals (van Fraassen 2002, p. 88, 224).

The Empirical Stance defends (i) by drawing on an idea by William James (1956). According to James, we have two central goals in our epistemic life: to believe as many truths as possible, and to believe as few falsehoods as possible. Since we cannot maximise both goals at once, each one of us implicitly or explicitly fixes their respective 'risk-quotients'. Each one of us must choose which of the two goals is more important (either in general or in specific contexts). Deductive logic and the theory of probability do not tell us how to make this choice. Our choice must therefore be based upon VEPPs (p. 87), which brings us back to the stances, but with a new twist. Stances now turn out to be important not just as renderings of some philosophical positions; they also turn out to be significant in how we organise our epistemic practices.

The Empirical Stance argues (ii) by revisiting the issue of scientific revolutions. Van Fraassen agrees with Kuhn's and Feyerabend's thought that, from the perspective of the pre-revolutionary old paradigm, the post-revolutionary new paradigm seems 'literally absurd, incoherent, obviously false, or worse–meaningless, unintelligible'. And yet, van Fraassen claims to deviate from Feyerabend by allowing for a different perspective *after* the revolution. From the post-revolution perspective, the pre-revolution viewpoint can be understood as a partial truth (p. 71). For instance, it follows from Einstein's Special Theory of Relativity that Newton's Laws of Motion are true for entities whose velocity is small when compared with the speed of light (p. 115) Moreover, van Fraassen thinks that scientific revolutions often result in the discovery of *ambiguities* in the old paradigm. Thus Newtonians did not realise that mass could be characterised as 'proper mass', 'gravitational mass' and 'inertial mass'. And they therefore regarded as absurd the notion that mass varies with velocity (p. 113).

According to ES, the litmus test for every epistemology is whether it can preserve the rationality of scientific revolutions while acknowledging the element of 'conversion' at their very heart. Objectifying epistemologies that describe and explain how our cognitive apparatus fits into the world, do not pass muster. They fail the litmus test since they invariably are enmeshed with the scientific theories of their day. The objectifying epistemology *en vogue* during the reign of the old paradigm licenses the old paradigm's epistemic ways. It therefore cannot but reject the epistemic practices of the new paradigm as irrational (p. 81).

Epistemic voluntarism does better for three reasons. First, it is prescriptive–evaluative rather than descriptive–explanatory (p. 82). This cuts the ties to prevalent scientific paradigms. Second, epistemic voluntarism is minimalist (cf. (i) noted earlier). And third, epistemic voluntarism gives emotion – or similar 'impulses' – a legitimate place in our epistemic life. Points two and three connect epistemic voluntarism to the stance-idea. Van Fraassen's thought seems to be that scientific paradigms are, or include, one or more scientific stances.

The Empirical Stance's example for how emotions can change one's epistemic options comes from Franz Kafka's short story *Metamorphosis*. One morning, Gregor, the son of the Samsa family, wakes up in the shape of a gigantic beetle, unable to communicate with humans. Initially his parents and sister, Grete, think of the beetle as their son or brother. Alas, this rendering of the situation makes their life unbearable. There is no way to maintain a normal family life when one family member is an insect. It is only when Grete eventually has an emotional breakdown that the parents find a way forward: they take the beetle for nothing but a beetle – and kill it (p. 106). Grete's emotion enables the family to recognise the situation more correctly, at least when judged retrospectively.

Going beyond van Fraassen's own words, we can use *Metamorphosis* also to illustrate another central claim of ES, to wit, the claim that scientific revolutions involve a re-interpretation of central rules guiding scientific work (chapter 4). The Samsa family, throughout the whole episode, operates with the rule *Protect the members of your family*. Initially this rule is used in a 'conservative way': Gregor and the beetle are taken to be the same person. And thus Gregor, the beetle, remains within the domain of the rule. After Grete's breakdown, however, the rule is interpreted in a 'revolutionary way': it is understood as legitimating the killing of the beetle. The family now thinks that the beetle has destroyed and replaced Gregor. Killing the insect secures Grete's well-being and takes revenge for Gregor. The change in the interpretation of the rule also involves the identification of an ambiguity: 'family member' as 'close relative that is human in outward appearance throughout their life,' and 'family member' as 'close relative regardless of their physical appearance'. The decision to kill the beetle meants that the parents and Grete reject the second meaning and come to insist on the first.

One central rule in empirical science is '*sola experientia*'. Defenders of the old paradigm use this rule in a conservative-defensive way. They insist that their paradigm is fully based upon experience (observation and experiment) and free of idle speculation. The proponents of the new paradigm instead accuse the old paradigm of violating *sola experientia*. They use *sola experientia* in a revolutionary way. For instance, Newton's critics identified his assumptions concerning absolute time and space as metaphysical baggage not licensed by experience. The upshot is that scientific revolutionaries do not simply throw scientific rationality overboard. But, they interpret it in radically new ways. And it needs an emotion-like 'impulse' to set off such developments (ibid).

The above-mentioned upshot is of course no more than a thumbnail-sketch of van Fraassen's epistemic voluntarism, but it is good enough for present purposes. To sum up: epistemic voluntarism emphasises the limited role of principles of rationality; distinguishes sharply between such principles and VEPPs; stresses the importance of VEPPs; and opposes objectifying epistemology. And van Fraassen regards epistemic voluntarism as the key to understanding the rational core of scientific revolutions.

5.5   Epistemic Voluntarism and Scientific Revolutions in AM

We can now return to the interpretation of AM. I shall discuss Feyerabend's position vis-à-vis the central ingredients of epistemic voluntarism.

5.5.1   *Objectifying or Voluntaristic Epistemology*

The Empirical Stance's main objection to objectifying epistemology is that the latter is unable to account for the rationality of scientific revolutions. An objectifying epistemology is always closely inter-twined with contemporaneous paradigms, and thus unable to conceive of their replacement as anything but irrational and absurd. How does AM stand vis-à-vis this argument?

To begin with, AM offers historical evidence for the claim that specific forms of epistemology are inseparable from contemporaneous paradigms, cosmologies or worldviews. This is a central point of AM's detailed interpretation of Galileo's controversy with the Aristotelians:

> Astronomy, physics, psychology, *epistemology*–all these disciplines collaborate in the Aristotelian philosophy to create a system that is coherent, rational and in agreement with the results of observation as can be seen from an examination of Aristotelian philosophy in the form in which it was developed by some mediaeval philosophers. (AM, p. 149; emphasis added)

Central in the Aristotelian epistemology-cum-psychology was a 'naïve realism with respect to motion', that is, the view according to which 'apparent motion is identical with real (absolute) motion . . .' and 'absolute motion is always noticed' (pp. 75, 90). In challenging naïve realism – for instance by introducing the telescope into astronomy – Galileo invented a 'new kind of experience'. Galileo had to persuade his contemporaries that the motion of the Earth was real even though it could not be observed. He tried to do so by reminding his readers of the fact that in some contexts they already accepted the 'relativity of motion': the motion of a painter's brush on a boat travelling through the Mediterranean could be identified relative to the canvas mounted on the easel, or relative to the surface of the Earth. By analogy, Galileo's readers were meant to allow that a stone falling from a tower likewise made at least three movements: towards the surface of the Earth, with the Earth around the Sun and with the rotating Earth (p. 86): 'It is this change which underlies the transition from the Aristotelian point of view to the epistemology of modern science' (p. 89).

Although Feyerabend agrees with van Fraassen that the forms of epistemology of a given time period T are typically closely intertwined with the scientific paradigms prevalent during T, there nevertheless is also an important difference between the two men regarding this issue. Other than van Fraassen, Feyerabend does not single out *one particular form* of epistemology – namely *objectifying* epistemology–as being especially prone to be so inter-twined. The position of AM seems to be that *all* forms of epistemology may end up becoming obsolete when paradigms or world-views change.

It is easy to see the plausibility of this view. In fact, one of van Fraassen's own analogies can help to make this case. *The Empirical Stance* suggests that epistemic voluntarism relates to objectifying epistemology as Carl von Clausewitz's *On War* (1832) relates to John Keegan's *The Face of Battle* (1976). The first offers general strategies for the conduct of any war; the second aims to capture, amongst other things, individual soldiers' experiences on the battlefield (at Agincourt, Waterloo and the Somme) (ES, p. 82). Van Fraassen's thought seems to be that whereas the descriptions offered in *The Face of Battle* no longer apply to soldiers' experiences today, Clausewitz's strategies still work. No doubt, there is some truth in this suggestion. And yet, there is also room for an objection. The development of warfare since the early nineteenth century – think of guerrilla warfare, popular uprisings, terrorism, nuclear, chemical and biological weapons – has made many pillars of Clausewitz's Hegelian philosophy or war obsolete (Wikipedia 2018). *Mutatis mutandis* for non-objectifying epistemologies: the mere fact that they do not involve many specific, and historically changing, psychological assumptions, does not protect them from becoming out-dated. Specific conceptions of coherence or the importance given to selected epistemic values might favour one paradigm P_1 over another paradigm P_2. But then, when P_1 is replaced by P_2, so might these conceptions of coherence and the importance given to certain epistemic values (cf. Kuhn 1977, and see later).

The last paragraph suggests that perhaps Feyerabend is right not to motivate a distinction between objectifying and non-objectifying epistemologies with reference to how closely they are intertwined with scientific theories. At the same time, it is worth emphasising however, that we find two kinds of epistemology in AM as well: epistemological anarchism, on the one hand, and all forms of 'rationalist' epistemologies, on the other hand. The latter hope to codify (scientific) rationality in a fixed set of rules or norms. Rationalists thereby expect to explain, predict, evaluate and prescribe scientists' choices. Karl Popper's 'critical rationalism' is of course Feyerabend's paradigm case of, and preferred whipping boy for, this form of epistemology. It is the main thesis of AM that all such epistemologies fail. Indeed, as Feyerabend writes concerning the Copernican hypothesis: it 'runs counter to almost every methodological rule one might care to think of today' (AM, p. 67).

Anarchist epistemology does not aim for a rigid system of methodological norms. Mindful of the failure of rationalist epistemology, anarchist epistemology is content to investigate successful scientific actors like Galileo, and to learn from their rhetoric and propaganda as much as from their scientific reasoning (as traditionally understood):

> Galileo ... exhibited a style, a sense of humour, an elasticity and elegance, and an awareness of the valuable weaknesses of human thinking, which has never been equalled ... Here is an almost inexhaustible source of material for methodological speculation and, much more importantly, for the recovery of those features of knowledge which not only inform, but which also delight us. (p. 161)

Ultimately, the commitment of anarchist epistemology is less to specific ideals of knowledge or science, and more to the aforementioned political or ethical ideals of humanitarianism and liberalism.

5.5.2   *Principles of Rationality and VEPPs*

Van Fraassen's epistemic voluntarism is committed to the view that principles of rationality do not tell us which scientific paradigms, or philosophical stances, we should adopt. Van Fraassen accepts a sharp distinction between universal stance-transcending principles of rationality and stance-or-paradigm-dependent values. Principles of rationality demand that our paradigms be consistent and probabilistically coherent. Such principles put limits on the number of contenders, but they typically leave more than one of them in the running. This is the leeway in which values must operate.

Feyerabend too recognises the importance of VEPPs in paradigm-choice. His case study on the Copernican Revolution repeatedly stresses Galileo's use of 'propaganda', 'psychological tricks', 'clever techniques of persuasion', 'emotion' and 'appeal to prejudices of all kinds' (AM, pp. 89, 81, 106, 143, 154). Feyerabend also claims that the ultimate success of heliocentrism depended on the emergence of 'new classes' for whom Copernicanism stood for values such as 'progress' and 'forward looking' (p. 210). This influence of VEPPs is not, for Feyerabend, something to be lamented. On the contrary, he regards VEPPs as essential for the progress of science: 'Copernicanism and other "rational" views exist today only because reason was overruled at some point in their past' (p. 155); or 'Without a frequent dismissal of reason, no progress' (p. 179). Scientific revolutionaries succeed because 'they do not permit themselves to be bound by "laws of reason", "standards of rationality" or "immutable laws of nature'" (p. 191). Feyerabend is even willing to speak of 'mob psychology' as a necessary feature in such revolutions (p. 211).

Two things need highlighting about these passages. The first is that VEPPs are here presented not as peacefully coexisting with principles of rationality, but as overriding them. I take this to be a by-product of Feyerabend's attempt to identify and emphasise VEPPs, and not a general thesis about the relationship between principles of rationality and VEPPs. The two can conflict, and they can cohere. The second thing to note about the passages quoted in the last paragraph is that they not only draw attention to situations in which VEPPs go against principles of rationality, but also suggest that such 'going against principles of rationality' might ultimately lead to a rational outcome. This suggests distinguishing between the *short term* and *long term*. The fact that scientists – under the influence of VEPPs – act irrationally *in the short term*, leads to rational results *in the long term*. In other words, it is the role of VEPPs to enable a fledgling scientific paradigm to grow and gain support until it has developed into a form that can live up to universal principles of rationality. Note that this picture of the role of rationality and VEPPs is at least *similar* to van Fraassen's view: after all, van Fraassen too highlights the need for emotions as enablers or triggers of paradigmatic change, a change that, when viewed retrospectively, leads to rational results.

Up to this point, I have presented Feyerabend's views as if for him principles of rationality were paradigm-transcendent – as they are for van Fraassen. But, this is not in fact how AM typically talks about rationality. It is more accurate to say that for Feyerabend principles of rationality are *internal* to paradigms or worldviews. This view surfaces when Feyerabend rails against 'universal laws, [and] universal ideas such as "Truth" [or] "Reason"'. He merely grants that 'it is often good policy to act *as if* such laws (such standards, such ideas) existed' (p. 189, emphasis added). He also writes that the 'idea of a . . . fixed theory of rationality rests on too naïve a view of man and his social surroundings' (p. 27) and that frequently 'a new physics or a new astronomy might have to be judged by a new theory of knowledge' (p. 153). Such judgements are difficult to square with van Fraassen's epistemic voluntarism.

This difference suggests distinguishing between van Fraassen's *moderate* and Feyerabend's *radical* epistemic voluntarism. According to the radical version, there are no permanently fixed principles of rationality; all there is, is several basic cognitive values, interacting with other values, and weighted differently in different worldviews, paradigms or stances. Intriguingly enough, van Fraassen himself notes this option when considering whether consistency really deserves the status of a stance-transcending demand of rationality. But, he firmly believes that the answer must be 'yes'. Someone who completely ignores consistency is unable to draw any distinctions (personal communication), and invariably fares badly in evolutionary history (van Fraassen 2004a, pp. 184–185). The radical epistemic voluntarist will demur. At issue is not whether consistency and coherence belong to our set of prima facie values; the issue is whether they should reign supreme. Graham Priest, for example, argues that consistency is a matter of degree and must always be weighed against other cognitive values such as 'simplicity', 'unity', 'explanatory power' or 'parsimony' (Priest 2005, p. 123).

If Priest is right about consistency and its relation to other epistemic values, then it becomes difficult for van Fraassen to maintain the distinction between *stance-transcending* principles of rationality and *stance-dependent* VEPPs. If consistency can be rationally overruled, then it cannot be the universal and necessary criterion of rationality. If consistency can be overruled, then a stance, which does so is not per se irrational. On this alternative picture, differences between rationally acceptable stances may be differences in what weight these stances give different 'cognitive values', including the value of consistency.

It should be added that the challenge to van Fraassen's emphasis on consistency does not just come from Priest's controversial views. Kuhn (1977) argues along similar lines. Kuhn offers 'shared epistemic values' (accuracy, consistency, scope, simplicity, fruitfulness) as the rational backbone of theory or paradigm-choice. But, Kuhn also insists that different scientists may rationally favour some values over others; interpret a given value in differently; or resolve conflicts between these epistemic values in variant ways (Lipton 2004, pp. 153–155). This is clearly in line with epistemic voluntarism. Note, however, that consistency is again part of the value-mix and not standing outside as the ultimate touchstone or arbiter.

All this suggests radical epistemic voluntarism. Rationality consists in one's honouring all or some of the epistemic values. Kuhn and Priest list some of these values, but no doubt there are more. Indeed, which epistemic values are can only be determined by research in cognitive psychology and the history and philosophy of science (including epistemology). This does not give us a firm and fixed base; but perhaps it is the *conditio humana* to cope without such foundation.

Feyerabend is clearly with Priest and Kuhn. Admittedly, some passages sometimes ring otherwise: 'Copernicanism and other "rational" views exist today only because reason was overruled at some point in their past' (AM, p. 155). Here 'reason' is not historicised or factorised into different cognitive values, but differently weighted by different communities. But, I take this sentence to be no more than shorthand; after all the same sentence puts inverted commas around 'rational'. Not to forget Feyerabend's granting that 'it is often good policy to act as if such laws (such standards, such ideas) existed'. In other words, couching some central tenets of AM in moderate garb makes the argument more palatable to readers unfamiliar with, or hostile to, the anarchist message. Moreover, and specifically with respect to consistency, remember that AM extensively rails against the so-called consistency condition, that is, the idea that a new scientific theory must be consistent with existing theories or observations (AM, p. 35). *Against Method* also proclaims that 'there is not a single science, or other form of life that is useful, progressive as well as in agreement with logical demands' (AM, p. 258–259).

To sum up: Feyerabend is a *radical* epistemic voluntarist; van Fraassen is a *moderate* epistemic voluntarist. Although the debate between these two positions can hardly be decided in a single subsection of a single paper, at least this much seems to me to be clear: there is at least a prima facie case in favour of radicalism.

5.5.3   *Incommensurability*

For van Fraassen scientific revolutions always involve *incommensurable* paradigms. Feyerabend is less adamant on this point. Although he writes of the 'Copernican Revolution' or Galileo's 'revolutionary' way of understanding motion, Feyerabend also insists that 'I never assumed that Ptolomy and Copernicus are incommensurable. They are not' (AM, pp. 23, 95, 114, 305). It is hard to know what to make of this *en passant* comment. Feyerabend and van Fraassen agree that scientific revolutions are cases where the newly suggested theory or paradigm appears 'absurd' (EM, pp. 71, 113; AM, pp. 64, 66, 81, 189). They also agree that in retrospect the paradigm change appears rational. And even though Feyerabend does not think Ptolomy and Copernicus were incommensurable, he is happy to speak of incommensurability between 'classical mechanics (interpreted realistically) and quantum mechanics (interpreted in accordance with the views of Niels Bohr), or between Newtonian mechanics (interpreted realistically) and the general theory of relativity (also interpreted realistically)' (p. 271). Why then does AM deny that Ptolemy and Copernicus are incommensurable? Two reasons may be important here. First, note the emphasis on '*interpretation*' in the last quote. Feyerabend maintains that two theories are commensurable or incommensurable *only under an interpretation*. For instance, he writes that 'instrumentalism ... makes commensurable all those theories which are related to the same observations language and are interpreted on its basis' (p. 279). Thus Ptolemy and Copernicus may well be incommensurable under the interpretations of seventeenth-century Aristotelians or Galileo, but well-nigh commensurable under the interpretation of, say, Tycho Brahe.

Feyerabend may also have a second reason for denying that Ptolemy and Copernicus are incommensurable. This is that for him the requirements for incommensurability are high. Two paradigms, A and B, are incommensurable when their respective facts

> cannot be put side by side, not even in memory: presenting B-facts means suspending principles assumed in the construction of A-facts. All we can do is draw B-pictures of A-facts in B, or introduce B-statements of A-facts into B. We cannot use A-statements of A-facts in B. Nor is it possible to *translate* language A into language B. (p. 270–271)

It is not obvious that these conditions are met in the clash of Galileo versus the Aristotelian philosophers. After all, Galileo was very much able to put geo- and heliocentristic assumptions 'side by side'. He would hardly have been able to write his 'Dialogue on the Two World Systems' otherwise.

It seems to me that this difference between van Fraassen and Feyerabend is important. It is of systematic interest to distinguish between revolutions in thought that involve *semantic* incommensurability in the strict sense outlined by van Fraassen, and revolutions in thought that do not. It is then an open question whether a given scientific revolution falls into the one or the other category, and what effects this has on the course of scientific debate.

5.5.4   *Absurdity and the Idealisation of Instantaneous Change*

Van Fraassen's account of scientific revolutions veers towards the instantaneous: problems mount for the old paradigm; scientists increasingly doubt that they can make it work; they fall into epistemic despair; they encounter a new, absurd, alternative paradigm; and then an emotion-like impulse enables them to reconceptualise this alternative as rational. Feyerabend's rendering is different. *Against Method* (and later writings) analyses the *extended process* of working out a new paradigm. Interestingly enough, van Fraassen sometimes expresses sympathies for Feyerabend's focus, but these sympathies do not influence van Fraassen's theory (van Fraassen 2000).

Feyerabend describes the interval between the reigns of successive paradigms as a time in which increasing numbers of scientists are willing to support a view that they recognise as 'inconsistent with ... plain and obvious [facts]' and with well-established theories (AM, p. 56). It takes 'sheer pig-headedness' (p. 155) for these scientists to develop an alternative: this is because every step in the direction of the new view produces 'nonsense' by the lights of received theories. Developing alternatives calls for several tools: prominent amongst them are ad hoc hypotheses; the accusation that the received evidence is 'contaminated' by theories that are false; the creation of new sensations with the help of new instruments; or the reinterpretation of what can be experienced (pp. 67, 89, 97, 99). In short: 'the 'irrationality' of the transition period [is] overcome ... by the determined production of non-sense until the material produced is rich enough to permit the rebels to reveal, and every-one else to recognise, new universal principles' (p. 270).

Although Feyerabend and van Fraassen differ in how much detail they provide on the 'transition period' from old to new paradigms, van Fraassen agrees with Feyerabend's insistence that to understand such periods calls for an understanding of ordinary (scientific) language as fluid, flexible and full of ambiguities. I have emphasised this aspect concerning *Metamorphosis* above. Suffice it to say that van Fraassen finds the treatment of language in chapter 1 of *Conquest of Abundance* (1999) particularly insightful. Here Feyerabend describes a radical conceptual shift in the Homeric world: from thinking that to be honourable *is* to be honoured by one's society, to thinking that you can be honourable even when your society fails to honour you. Initially the shift was hard to make sense of. When Achilles first drew the distinction, his Greek audience, as Homer puts it, 'fell silent, for he had spoken in stunning ways' (Feyerabend 1999, p. 19). How could Achilles convince the Greeks? As Feyerabend has it, by exploiting the openness of language and the ordinary speakers' temporary tolerance of incomprehension: 'we ... find that ordinary people ... readily accept statements, which sound strange to their neighbours and nonsensical to scholars' (p. 32). Given the right incentives or interests, we are willing to pick up and use expressions that become meaningful only when accompanied by further new expressions (that initially were just as 'stunning'). What is more, language is full of ambiguities and analogies that can be exploited by clever and determined men like Achilles or Galileo. Galileo's parallel between the painter in the boat and human on earth convinced people that earthly movement might remain unrecognised. And Achilles' suggestion that being honoured by the Gods doesn't imply being honoured by one's communities paved the way to breaking the link between being honourable and being honoured by one's own society (p. 36). Feyerabend sees these linguistic techniques at work regardless of whether the given 'revolution' involved incommensurability or not (p. 270).

5.5.5   *Impulses and Rules in Scientific Revolutions*

Van Fraassen takes emotions, or emotion-like 'impulses', to be the necessary preconditions for scientific revolutions. Such emotions and impulses have the potential to bring about a radical restructuring of choice-situations: they generate the new sets of 'live options'. Unfortunately, van Fraassen remains a bit sketchy on the details, relying primarily on the intuitive plausibility of the Samsa family's radical change in view concerning the beetle in Gregor's room.

To get clearer on the range of possible positions concerning this issue, it seems useful to bring in a third voice, in addition to van Fraassen and Feyerabend: Ernan McMullin (2007). McMullin is happy to admit that Copernicus' or Galileo's positions seemed 'absurd' to their Aristotelian opponents (McMullin 2007, p. 172). But, he denies that the available sources on either Copernicus or Galileo reveals the 'epistemic despair' or 'Sartrean moment' that van Fraassen's impulse theory of scientific revolutions requires. Copernicus arrived at his position via a 'mathematical reworking of the traditional Ptolemaic data', and when Galileo 'came along, a rational bridgehead was already there'. McMullin acknowledges that both Galileo and his opponents 'could claim to be rational in choosing the position [they] did' (pp. 172–173). The core of van Fraassen's reply is that the need for an emotion-like impulse is a 'point of logic': some such impulse *must be* operative if rational actors are to switch to a position they had regarded as absurd (van Fraassen 2004a).

The similarities and differences between McMullin, van Fraassen and Feyerabend are intriguing. McMullin and Feyerabend share an emphasis on the Copernican Revolution as a long, drawn-out process. As we saw in the last section, van Fraassen's idealised and simplified account sometimes sounds as if scientific revolutions were instantaneous events.

In another respect, Feyerabend is closer to van Fraassen than to McMullin. For Feyerabend and van Fraassen the switch from geo- to heliocentrism was not a switch governed solely by established rules of scientific method plus theory-neutral evidence. I have already quoted several passages from AM that stress the importance of values, biases, rule breaking and prejudice. Incidentally, even McMullin admits that good data and the scientific method were not quite enough to push Copernicus towards heliocentrism. Copernicus was also influenced by the idea that 'the sun, the source of light and life [belongs] at the centre of the universe' (McMullin 2007, p. 172). Be this as it may, Feyerabend's emphasis on values and impulses clearly puts him closer to van Fraassen than McMullin.

And yet, although Feyerabend and van Fraassen agree on the importance of values and impulses, they have rather different conceptions of what these impulses are. In van Fraassen's case, McMullin seems right when he speaks of 'Sartrean moments': these moments are closely modelled on religious or existential conversions *of individuals*. Feyerabend's impulses, by contrast, are collective responses to changing *social–political conditions*: it was, he writes, the rise of 'a new secular class' that was crucial for the ultimate success of Copernicanism. The new secular class was opposed to 'the barbaric Latin spoken by the [Aristotelian] scholars', to 'the intellectual squalor of academic science', to its 'uselessness' and to 'its connection with the Church'. Copernicus came to represent 'progress', 'the classical times of Plato and Cicero' and 'a free and pluralistic society'. Galileo exploited this setting 'and amplifie[d] it by tricks, jokes, and *non-sequiturs* of his own' (AM, p. 154). Feyerabend does not suggest that these political interests by themselves constituted scientific arguments against Aristotelianism. He rather sees them as elements that weakened 'the influence' of the Aristotelian considerations. People at the time paid less attention to these considerations; and they were willing to bet on Galileo's heliocentrism even while recognising that the view was, in the light of their Aristotelian background, absurd.

We do not, of course, have to see collective social responses and individuals' 'Sartrean moments' or 'epistemic despair' as excluding one another. No doubt, both can occur together. And yet, historical scholarship on the Copernican Revolution, especially by Mario Biagioli (1993, 2006), suggests that social-political issues go a very long way in explaining the actions of Galileo, his supporters and his opponents (cf. Finocchiaro 2001). It is much more difficult to discern from Galileo's writings when and where epistemic despair played a role. Van Fraassen agrees as much in suggestion that the role of emotion is more a point of 'logic' than a point of the historical record.

Finally, there is also a clear respect in which van Fraassen and McMullin are closer to each other than either of them is to Feyerabend. I am referring to van Fraassen's and McMullin's attempts to save the rationality of scientific revolutions. In van Fraassen's case, this takes the following form. Principles like '*sola experientia*' continue to be followed throughout scientific revolutions, albeit that – under the influence of different 'impulses' – they are interpreted fundamentally differently: *conservatively* in defence of an existing paradigm; in *revolutionary fashion* in defence of a new paradigm. Van Fraassen's discussion of '*sola experientia*' builds directly upon Feyerabend's paper 'Classical Empiricism' (1970d). Van Fraassen is particularly intrigued by Feyerabend's suggestion that the '*sola experientia*' of the natural philosophers of the seventeenth century played a similar role as did the Protestants' '*sola scriptura*'. Nevertheless, it is striking that van Fraassen's and Feyerabend's take on '*sola experientia*' is very different. Feyerabend is eager to establish that the rule is 'vacuous' (Feyerabend 1970d, p. 41). It is vacuous because it gives advice only in a setting in which people already agree on the importance of experience; agree on what counts as experience; agree on how to interpret experience; and agree on what can be derived from a given experience. But if all this is already settled, then '*sola experientia*' does not provide any new advice; it merely 'reinforce[s] an already existed faith' (40); it is no more than a 'party line [..]' (51). This is directly parallel to the Protestants' '*sola scriptura*'. The latter too only makes sense if the extension of 'scripture', its interpretation, and its principles of guidance have already been agreed in a community. In both cases the 'rejection of authority . . . does not lead to a more critical attitude. It leads to the enthroning of new authorities . . . : scripture on the one side, experience on the other' (p. 50).

Feyerabend is particularly interested in the 'classical empiricism' of Newton. To cut a long story short, Feyerabend highlights Newton's strategy for marshalling evidence for his views. First, an idea is made familiar by being repeated and illustrated; and second, the familiarity thus established is then used 'as if it were an additional source of support'. For Feyerabend this strategy is 'not different from political propaganda' (p. 51). This observation sounds like a criticism of Newton's strategy, but Feyerabend does not mean it in this way. On the contrary, he concludes that Newton's circular way of reasoning is 'democratic' in that the same technique can be used by anyone and applied to just 'any idea'. And this is good for scientific pluralism and a democratic and humanitarian science.

The difference between van Fraassen and Feyerabend is stark at this point. Van Fraassen wants to use '*sola scriptura*' to secure empiricist rationality across scientific revolutions. Feyerabend denies that the principle captures any form of rationality, empiricist or otherwise. He tries to convince us that it is no more than a vacuous party line, 'an ornament', potentially leading to a positive, pluralist outcome. Van Fraassen wants to save a continuity of rationality through revolutions; Feyerabend seeks to analyse the interplay of scientific aspirations, social circumstances, rhetoric and power.

5.6 Relativism in van Fraassen and Feyerabend

In §5.2, I contrasted van Fraassen's *moderate* with Feyerabend's *radical* version of epistemic voluntarism. The moderate version draws a sharp distinction between principles of rationality and VEPPs. The radical version, by contrast, allows for no permanently fixed principles of rationality; instead there are basic cognitive values, interacting with other values, and weighted differently in different worldviews, paradigms or stances. In this last section, I want to highlight the different implications of the two positions regarding epistemic relativism.

Arguably even van Fraassen's moderate epistemic voluntarism is committed to a form of epistemic relativism. It consists of the following theses:

1. The epistemic status of judgements is relative to stances.
2. Different stances evaluate the same judgements differently.
3. There is no perspective from which stances can be neutrally and absolutely ranked.
4. The move from one stance to another can have the character of a 'conversion:' principles of rationality combined with empirical data cannot compel a transition from one stance to another.

Van Fraassen himself invokes the idea of conversion as follows:

> Being or becoming an empiricist will then be similar or analogous to conversion to a cause, a religion, an ideology, to capitalism or to socialism, to a worldview such as Dawkins's selfish gene view or the view Russell expressed in 'Why I am Not a Christian'. (van Fraassen 2002, p. 60)

Van Fraassen is ready to admit the relativistic implications of epistemic voluntarism: 'If this is relativism, it is certainly not debilitating relativism—it is only an acknowledgement of the logic of this aspect of the human condition' (van Fraassen 2004b, p. 11). I take it that by 'debilitating relativism' van Fraassen means a form of relativism that makes its advocate unable to judge or argue. Perhaps he is thinking of versions of epistemic relativism that declare all stances to be 'equally valid'. Clearly, if all stances are equally valid then there cannot be much point in arguing about them.

Note also that van Fraassen's relativism leaves room for rejecting some alternative epistemic practices as *absolutely* irrational: after all irrationality can be measured against a unique set of principles of rationality (of deductive logic, the theory of probability and the practical syllogism). Relativism comes into its own only when two epistemic practices differ only in their VEPPs. In such cases, the assessment of alternative epistemic practices is based on stance-internal and potentially stance-specific epistemic and other values.

Clearly, the epistemic relativism of AM is more radical in precisely the way in which radical epistemic voluntarism is more radical than moderate epistemic voluntarism. The epistemic relativism of AM goes 'all the way down' – there are no universally valid principles of rationality. Feyerabend rejects 'universal standards, universal laws [and] universal ideas' (AM, p. 189). Still, his relativism does not prevent him from making evaluative statements about his own and other frameworks. Feyerabend regards it as *sceptical rather than relativist* to claim that 'every view [is] equally good, or [is] equally bad' (p. 189). And AM is happy to say that '[we] can say today that Galileo was on the right track' (pp. 26, 155). Such judgements are not absolute of course but are based on our epistemic practice. While such passages show that Feyerabend, just like van Fraassen, rejects a debilitating 'equal-validity' relativism, there is nevertheless another 'equality thesis' central in AM: to bring about rational progress – as measured with local and contingent criteria – it always needs 'irrational means such as propaganda, emotion, ad hoc hypotheses, and appeal to prejudices of all kinds' (p. 154). Finally, AM is of course famous – or notorious – for denying that science is superior to 'myth, religion, magic [and] witchcraft' (pp. 291, 298). For Feyerabend science is just 'one of the many forms of thought that have been developed by man, and not necessarily the best' (p. 291). And the same is true for philosophy (p. 298).

5.7   Van Fraassen, Feyerabend, Boghossian

In this final section, I shall discuss how the two forms of relativism – the moderate relativism of ES, and the radical relativism of AM– fare against four anti-relativist arguments put forward by Paul Boghossian. This will help to further clarify the differences and commonalities between the two positions.

(i) By 'absolute relativism' Boghossian means a form of relativism that works with a mixture of absolute and relative principles. The paradigmatic case of this view is a relativism of manners based on one absolute principle: 'When in Rome do as the Romans do'. Or think of subjective Bayesians for whom the Bayesian formula is the one and only absolute principle (Boghossian 2011, p. 67). Boghossian rejects absolute relativism. By accepting the existence of one absolute principle, Boghossian submits, the relativist has lost what surely must be her strongest card, to wit, worries how absolute principles fit into the empirical world, and how they can be known by finite and fallible creatures. Moreover, the absolute relativist has no good answer to the question why there could not in principle be more than one absolute norm (p. 68).

Van Fraassen's moderate relativism–voluntarism is a clear instance of an 'absolute relativism'. After all, van Fraassen treats principles forbidding inconsistency and incoherence as definitive of rationality, and as different from VEPPs. Still, I suspect that he would be unmoved by Boghossian's attack. In allowing for absolute principles, van Fraassen and Boghossian are in the same boat; thus if van Fraassen owes us an account of how absolute principles can fit into the empirical world, so does Boghossian. Van Fraassen also has a suggestion concerning the question why there could not in principle be more than just a few absolute norms. 'There could be', he might well reply, 'but as the variation is stances and VEPPs shows, there aren't more than the ones I have identified'.

The Feyerabend of AM can give a different answer. He can simply deny being committed to absolute relativism. Feyerabend's radical relativism rejects the distinction between absolute and relative principles or values.

(ii) Boghossian readily acknowledges that our epistemic practices vary, but he denies that that this variation supports relativism. What variation is there which can be explained by the fact that our absolute rules are sometimes vague and unspecific? They leave room for choice. (Boghossian, personal communication; 2006, p. 110). This suggestion seems to fit with van Fraassen's *moderate* relativism with its principles of rationality that leave our choices of stances or paradigms underdetermined. This underdetermination is removed only once VEPPs do their work. Feyerabend's relativism deviates fundamentally from Boghossian's proposal.

Who is right? The first thing to note here is that Boghossian's idea does not, in fact, block relativism. If true, all it suggests is that the *scope* of relativism is not unlimited. But, the breadth of the scope remains completely open. Clearly, Boghossian, van Fraassen and Feyerabend are likely to have very different views on this breadth.

Boghossian is likely to press his case by arguing as follows. If we allow, as we should, for appropriate forms of idealisation and abstraction, then surely we will be able to construct general and absolute epistemic principles to which every normal human being is at least implicitly committed.

Feyerabend should not be overly impressed. Yes, indeed, he should say, we might proceed in the way Boghossian suggests. But, we should not expect this methodology to lead to one unique outcome. On the contrary, work done in this way is faced with all the old issues concerning the underdetermination of theory by observation. Moreover, it might well be highly artificial and contrived to bring all our epistemic folkways under one small set of absolute epistemic principles. And last, but not least, what should we do with the actors' own perspective on their epistemic folkways? Should we simply ignore this perspective? If not, what then should we say when the actors do not recognise their own reasoning in the epistemologists' reconstructions and idealisations?

(iii) Stances and perhaps even paradigms can be more or less different, more or less distant, from one another. The greater the difference or distance, the more we need the idea of 'conversion' for capturing what happens when the folk or scientists shift from one stance or paradigm to another. And it is only when conversion is needed for capturing the change that epistemic relativism is vindicated.

Parts of Boghossian's 2006 book can be read as offering a suggestion for how the relativism-motivating distance or difference between stances or paradigms can be captured. Boghossian distinguishes between 'fundamental' and 'derived' 'epistemic principles'. A *fundamental* principle concerning observation licenses perceptual beliefs under certain general conditions. A *derived* principle concerning observation licenses the perceptual beliefs of a specific person or perceptual belief given a specific instrument (like a microscope or telescope). Boghossian claims that two 'epistemic systems' – that is, two systems of epistemic principles – are 'fundamentally different' when they differ in at least one fundamental epistemic principle. And the fundamental difference of epistemic systems is what defines a relativistic setting. Of course, Boghossian's interest in all this is to bury relativism, not to praise it. He therefore goes on to argue that relativists have so far failed to offer a single convincing case of such fundamental difference between epistemic systems. In particular, Galileo and Cardinal Bellarmine did not differ over any fundamental epistemic principle (Boghossian 2006, pp. 63–69, 90–91, 103–105).

Can Boghossian's concepts and criticisms be applied to van Fraassen's and Feyerabend's relativism? Have they offered convincing examples of differences in fundamental epistemic principles? To my mind neither form of relativism is threatened by Boghossian's considerations. To begin with, it is most unlikely that Feyerabend or van Fraassen would accept Boghossian's criterion for a relativism-inducing difference instances, that is, a difference in at least one fundamental principle. Feyerabend's and van Fraassen's perspectives are coherentist rather than foundationalist. What distinguishes Cardinal Bellarmine from Galileo is not one fundamental epistemic principle but a whole host of beliefs and VEPPs. It is the number and weight of these differences that require a conversion, not the fundamental character of one of them. The distance from Boghossian increases further as we shift from moderate to radical epistemic relativism: the latter does not accept that everyone must share (the interpretation of) the same basic epistemic values.

(iv) Boghossian (2011, pp. 60–66) finds epistemic relativism inherently unstable. On the one hand, the relativist allows that epistemic systems fundamentally different from her own are, in some sense, as valid as her own. On the other hand, the relativist also prefers her own epistemic system and does not give it up. How can these two attitudes be reconciled? Boghossian is doubtful that relativism can deliver a plausible solution.

Van Fraassen's response is perhaps best captured in the following remark (which was not addressing Boghossian's considerations):

> I remain convinced that genuine, conscious reflection on alternative beliefs, orientations, values–in an open and undogmatic spirit–does not automatically undermine one's own commitments. (van Fraassen 2011, p. 156)

Of course, we need an argument defending this conviction. I submit that both the moderate and the radical relativist can provide such argument. Here is what Feyerabend might say. Under certain conditions, we can – from the perspective of our paradigm (or stance) – recognise the VEPPs and beliefs of another paradigm as justifiable. That is, we can come to see the VEPPs and beliefs of another paradigm as rational provided only that we can identify a way *of justifying them with reference to some plausible combination and weighting of epistemic values*. If this proves possible, then the other paradigm is in some sense 'equal' to our own. And yet, the fact that we can see the other paradigm in this light does not give us a reason to convert to it. After all, we might well have VEPPs and beliefs that differ from those of the other paradigm. And our VEPPs and beliefs might give us sufficient reason not to convert. Van Fraassen would agree, except for the words in italics. He would say that in justifying the VEPPs and beliefs of another paradigm, we can also invoke shared stance-transcending principles of rationality. Be this as it may, neither Feyerabend nor van Fraassen is threatened by Boghossian's argument.

5.8 Conclusion

In this chapter, I have compared and contrasted three themes in van Fraassen's ES – the idea of stances, epistemic voluntarism, relativism – with motifs and arguments in Feyerabend's AM. I hope to have shown that epistemological anarchism is naturally read as a stance; that van Fraassen deviates from Feyerabend in defending a *moderate* rather than *radical* version of epistemic voluntarism; that the two philosophers differ in their understanding of scientific revolutions (e.g. Feyerabend's rendering of scientific revolutions is less individualistic, less instantaneous and less concerned with 'saving their rationality' than van Fraassen's account); and that the difference in voluntarism reflects itself in two versions of epistemic relativism. I concluded by arguing that both versions have the argumentative resources to defend themselves against Boghossian's anti-relativist arguments.

CHAPTER 6

Feyerabend Never Was an Eliminative Materialist
Feyerabend's Meta-Philosophy and the Mind–Body Problem

Jamie Shaw

6.1 Preamble

David Chalmers famously outlined the 'hard problem of consciousness' which, in essence, involves discerning the relationship between material and mental processes. The hard problem, it is often thought, is a distinctively philosophical problem for which no empirical answer is possible:

> To explain experience, we need a new approach. The usual explanatory methods of cognitive science and neuroscience do not suffice. These methods have been developed precisely to explain the performance of cognitive functions, and they do a good job of it. But as these methods stand, they are only equipped to explain the performance of functions. When it comes to the hard problem, the standard approach has nothing to say. (Chalmers 1995, p. 204)

This formulation of the mind–body problem, as Kim (1998) points out, dominated the later part of the twentieth century. Eliminative materialism provides one solution, which states that, ontologically speaking, there are only material processes. Feyerabend is often cited as an arch-defender of this view. A closer analysis, though, shows that this isn't the case. Rather, Feyerabend's resolution of the mind–body problem lies at a deeper level. Specifically, he is concerned with how philosophical paradoxes, including the mind–body problem, are understood. For Feyerabend, there are many possible answers to the mind–body problem that come from stipulating ontological bases that can be subjected to empirical scrutiny. The empirical scrutiny leads to progress, not the 'solution'.[1] Feyerabend endorses the

---

[1] This conclusion is not unique to Feyerabend. Place (1970), Patricia Churchland (2002), Dennett (2006) and others also construe materialism as a scientific hypothesis. Thus Bickle (1998, p. 40) is wrong to suggest that 'traditional' formulations of the mind–body problem were ontological and reformulating the debate as scientific is a 'recent' development; this formulation was there from the beginning.

opposite view from Chalmers: the hard problem is easy, and the easy problems are hard.

To complicate things, Feyerabend changes his mind about what makes for a genuine alternative theory. In the 1960s, Feyerabend follows Popper: alternatives have different empirical content. In the 1980s/1990s, Feyerabend follows Aristotle: alternatives have differences for how to live. The result of this discussion is to lend greater clarity to an under-discussed feature of Feyerabend's thought and outline a position worthy of further consideration.

Eliminative materialism, in its most recognised form, was articulated and defended by the Churchlands. Feyerabend is cited as a primary influence on their views (P. M. Churchland 1992; 1997; P. S. Churchland 1988). But there are many ways eliminative materialism is cast; one is Feyerabendian while the others are not. One states that eliminative materialism contains the ontological thesis that only material, not mental, entities or processes exist. 'Eliminative materialism is the thesis that our common-sense conception of psychological phenomena constitutes a radically false theory, a theory so fundamentally defective that both the principles and the ontology of that theory will eventually be displaced, rather than smoothly reduced, by completed neuroscience' (Churchland 1981, p. 67). Feyerabend, I claim, never held this view. Pace Keeley (2006, p. 11), he was not a 'metaphysical materialist'.

A weaker version of eliminative materialism states that postulated mental categories *may be* false and *could be* reduced to material processes in some future (successful) scientific theory.[2] They can be eliminated from our ontology as science progresses like 'ghosts' or 'humours' were. Feyerabend supported this view its consequence: *alternatives* to eliminative materialism are also possibly true. Within this camp, materialists often provide *predictions* of which theories will be accepted in future science. Rorty, for example, argues that 'materialism embodied many confusions, but at its heart was the unconfused prediction about future empirical inquiry, which is the Identity Theory' (Rorty 1965, p. 54).[3] Churchland similarly declares that 'By the late sixties, every good materialist expected that epistemological theory would

---

[2] It is common to speak of mental processes reducing to *brain* processes. However, mental processes may reduce to other processes, as embodied or distributed cognition theories suggest. Since materialism is compatible with both reductions, I will speak of the reduction to material processes to be neutral on the location of mental correlates.

[3] Rorty's own view on this is obscure, since he also writes that 'there is simply no such thing as a method of classifying linguistic expressions that has results guaranteed to remain intact despite the results of future empirical inquiry' (Rorty 1965, p. 25) which, presumably, includes materialism.

one day make explanatory contact, perhaps even reductive contact, with a proper theory of brain function' (Churchland 1992, p. 30). Dennett disagrees about what will be accepted in the future, but agrees that scientific progress will provide the solution. He and Churchland 'differ only over where the main chance for progress lies' (Dennett 2006, p. 193). In Churchland's case, he justifies the view that folk psychology is not only elimin*able* but should be elimin*ated*, on Lakatosian grounds:

> The history [of folk psychology (FP)] is one of retreat, infertility, and decadence ... Both the content and success of [folk psychology] have not advanced sensibly in two or three thousand years. The FP of the Greeks is essentially the FP we use today. (Churchland 1981, pp. 7–8)[4]

This lack of progress is a problem given that, according to Churchland, FP cannot account for a wide variety of phenomena (e.g. sleep, mental illness, hallucinations, learning, etc.) and has made no attempts to show its relationship to neighbouring sciences.

Feyerabend would not and, as I argued elsewhere, *could not* predict which theory would be accepted in the future (Shaw 2018a; 2018b). Moreover, he would reject Churchland's elimination of FP. His response to FP's problems would be: 'Too bad; try again!' (Feyerabend 1962b, p. 155). These problems could have led Churchland to be its arch-defender of FP by trying to overcome its difficulties. Feyerabend's pluralism dictates that we shouldn't eliminate any theory, regardless of its faults, and we should maintain FP alongside materialism (see Shaw 2017). It would also have been quite embarrassing for Feyerabend to predict that materialism would be true and reject it later in his career. While this difference is important, the common basis of the weaker view is shared by Feyerabend and others: the viability of materialism is determined by empirical scrutiny.

In Section 2, I outline Feyerabend's meta-philosophy. Specifically, I show his understanding of the goals of philosophy. In Section 3, I show how Feyerabend applies this meta-philosophy to the mind–body problem. The result is that there is no 'problem' in the traditional sense and the paradox can be resolved in many ways. In Section 4, I provide two other arguments Feyerabend provides in defence of materialism. Finally, in Section 5, I analyse Feyerabend's change of mind concerning how the mind–body problem should be resolved. I show how this change of mind is underpinned by Feyerabend's defence of 'Aristotle's principle'.

---

[4] For an up-to-date, more optimistic and nuanced empirical assessment, see Mendelovici (2018).

6.2   Feyerabend's Meta-Philosophy

In one of his earliest papers, Feyerabend considers the implications of the 'paradox of analysis' for the nature of philosophy (Feyerabend 1956). The paradox states that if a proposition, p is equivalent to q (i.e. the biconditional p if q is uninformative), then the discovery that $p = q$ is trivial since it 'does not convey any knowledge which one does not already possess' (p. 93). This is only a paradox, Feyerabend contends, if triviality is understood as a semantic notion, making it a feature of a language, rather than an agents' understanding of p or q (the 'pragmatic' understanding of triviality). This formulation is grounded in the convictions that (a) philosophy has no subject matter of its own and (b) can make progress (i.e. gains in knowledge). (a) must be true since if triviality is understood pragmatically, then 'discoveries' would be descriptive accounts of language learning. Rather, proponents of (a) claim that the domain of philosophy is the conceptual analysis of statements.[5] (b) must be the case or else there would be no paradox. Therefore, Feyerabend concludes, the paradox of analysis entails that '*philosophy cannot be analytic and scientific*, i.e. interesting, progressive, about a certain subject matter, informative *at the same time*' (p. 95).

Feyerabend's meta-philosophy is further developed from his engagement with Wittgenstein's *Philosophical Investigations* (PI). He was well acquainted with Wittgenstein's work during his Ph.D.,[6] and wrote a tortuously complicated paper on the *Philosophical Investigations* (Feyerabend 1955).[7] According to Feyerabend, the primary tasks of the *PI* are: (a) elaborate a theory of meaning, T, which claims that 'Every word has a meaning. This meaning is correlated with the word. It is the object for which the word stands' (Wittgenstein 1953, p. 1), (b) criticise this theory, and (c) offer his own theory of meaning, T'. T' is the view that 'the meaning of the elements of a language-game emerges from their *use*' (p. 123). For (c), Feyerabend claims that he will 'state what seems to be Wittgenstein's own position on the issue. This position will be formulated as a philosophical theory, T', without implying that Wittgenstein intended to develop a philosophical theory (he did not)' (Feyerabend 1955, p. 99). Wittgenstein's criticisms of T are well known and do not require attention

---

[5] This was, famously, the view of many of the logical empiricists. For a discussion of this as it pertains to the mind–body problem, see Feigl (1950).

[6] See Stadler (2010) and Collodel (2016).

[7] Future thinkers (J. L. Austin, Norman Malcolm, J. O. Urmson, G. J. Warnock and Stanley Cavell) reinvigorated Wittgenstein's quietism within the 'ordinary language' movement. Because they offered new arguments, Feyerabend addresses them separately (see Feyerabend 1963a; 1969a).

here. What is important here is that he wants to criticise it. On this point, Feyerabend writes:

> Wittgenstein does criticise; but his criticism is of a particular kind. It is not the kind of criticisms which is directed, for example, against a wrong mathematical calculation. In the latter case the result of the criticism is that a certain sentence is replaced by its negation or by a different sentence ... For him 'the results of philosophy are the uncovering of one or another piece of plain nonsense and of bumps that the understand has got by running its head against the limits of language' (119), and his aim is 'to teach you to pass from a piece of disguised nonsense to something that is patent nonsense' (464) and in this way to clear up 'the ground of language' (119). But that can only mean that 'the philosophical problems should *completely* disappear' (133); for if the aim has been reached, 'everything lies open to view and there is nothing to explain' (126). This implies that the formulation of T' as used within the critical procedure cannot be interpreted as a new theory of meaning, for it is applied with the intention of making the language-games ... 'lie open to view' ... without any question arising as to how it is that words become meaningful. (Feyerabend 1955, p. 126)

The idea here is that Wittgenstein's criticisms cannot be that T is *false*, and T' is *true*, but that the formulation of philosophical theories is misguided in the first place. Philosophical theories, to use the language of the *Tractatus*, are ladders to be thrown away once they are understood. Philosophical theories *improperly* understood, are a 'house of cards' according to Wittgenstein, that collapse once we learn how to play a language-game. Feyerabend claims that Wittgenstein must assume that 'there is a *sharp line* between the 'house of cards' on the one hand and the language-games on which they are built on the other' (p. 127).[8] But if philosophical theories are just language-games, then

> one may ask why Wittgenstein tries to eliminate this theory as well as other philosophical theories ... Nevertheless Wittgenstein tries to eliminate this theory as well as other philosophical theories. But this attempt can only be justified by assuming that there is a difference between using a sign (playing a language-game) and proceeding according to theory *T*. The procedures which are connected with theory *T* are supposed to not to be taken as parts of a language-game, they constitute a sham-game which is to be destroyed. How is this attitude to be understood? (ibid).

Feyerabend argues that Wittgenstein maintains this view because he retains, unwittingly, the conception of philosophical theories from the

---

[8] This point comes up again in a similar context (Feyerabend 1969a, p. 148).

Tractatus: 'The word 'philosophy' must mean something which stands above or below, not beside the natural sciences' (4.III) (replacing 'natural sciences' with 'language-games'). If philosophical theories are language-games, then there is no asymmetry between eliminating them over ordinary idioms. Feyerabend rejects this conception of philosophy and Wittgenstein's quietism (do not construct philosophical theories). Instead, Feyerabend opts for Tarski's distinction between object-languages and meta-languages where philosophical theories are meta-languages that 'do not necessarily disturb the language games they are supposed to describe' (Feyerabend 1955, p. 128). Feyerabend claims that 'this solution would not agree with Wittgenstein's, but it would retain several elements of his philosophy: (1) his criticisms of *T*; (2) his statement of *T*; (3) his observation that language-games may be disturbed by other language-games' (p. 129).

What can we infer about Feyerabend's meta-philosophy from this? Recall that the paradox of analysis entails that philosophy must be either scientific (i.e. has an empirical domain) or analytic (i.e. analyses statements).[9] From Wittgenstein, Feyerabend claims that analytic philosophy qua T (i.e. philosophy that aims at 'complete clarity') is *impossible*. As Preston notes, 'throughout his career Feyerabend railed against analytical and 'linguistic' philosophy, insisting that philosophy should help science and humanity progress, rather than hinder them or simply try to clarify science from the sidelines' (Preston 1997a, p. 11). But it isn't just the normative demand that philosophy should aid in progress, but that analytic philosophy qua conceptual clarification is incoherent. Analytic philosophy is a language-game that *describes* other language-games. This makes it a branch of empirical linguistics. This entails that there isn't a 'choice' between analytic and scientific philosophy; philosophy can *only be* scientific.

These conclusions are influential on Feyerabend's later thinking. In 'Problems of Empiricism', for example, he declares that epistemology should be a part of scientific research:

> the question as to what is more fundamental-sensation or thought, observation or theory-is generally assumed to be independent of scientific research ... Even worse, it is assumed that the relation between mental

---

[9] Note that this way of putting it sounds strange post-Kripke (1972) where we can have analytic a posteriori truths. But this observation should not affect Feyerabend's argument since he is concerned about the *subject matter* of philosophy which is untouched the epistemic–logical distinctions Kripke attempts to make ('Water is H_2O' is synthetic regardless, or whether or not it is necessarily true).

events (inside the human observer) and physical events (outside the observer) can be decided by *philosophical* argument, by the abstract consideration of possibilities, without any help from the exigencies of empirical research. (Feyerabend 1965a, p. 147)

In *Against Method*, Feyerabend argues that scientific methodologies contain empirical assumptions and are therefore liable to change as science changes (Feyerabend 1975/1988, p. 233). The same theme appears in Feyerabend's dispute with Popper about realism in quantum mechanics; realism should be rejected because it leads to a problematic empirical theory (see Kuby, this volume). He even warns his readers about missing this distinction: '[Popper's] model breaks down in a finite world, but the breakdown will never become visible to a philosopher who hides factual assumptions behind "logical" principles and "methodological" standards" (Feyerabend 1970e, p. 88). Feyerabend's unrelenting naturalism remains a (largely unspoken) backbone of his mature views. Thus, Preston is wrong to say that Feyerabend agreed with Popper that epistemology concerns third-world 'products' of knowledge (Preston 1997a, p. 12); Feyerabend was always a thoroughgoing naturalist and took epistemological questions to be part and parcel with science.

But there is something missing here. If philosophy is only scientific, what are we to make of his pronouncement that we should invent new theories? What grounds this admonishment if philosophy is scientific? This poses a deep question that Feyerabend never explicitly tackles: what is the source of norms apart from those, which appear in particular forms of life? Feyerabend wavers on this point. In *Science and a Free Society*, he bites the bullet and accepts that norms are only valid within a given form of life. This leads to relativism. Elsewhere, Feyerabend responds to this problem, indirectly, in an added footnote in the reprinted Wittgenstein paper:

> Wittgenstein's main point, viz. that principles, general ides, metaphysical dreams receive content through a practice, may be retained provided we permit the researcher to build up new practices and do not demand that he restrict himself to exiting practices entirely. (Feyerabend 1955, p. 128 fn. 21)

This idea here comes later in Feyerabend's corpus: understanding emerges from *contrast*, which *presupposes* pluralism. The process of learning a language-game comes from contrasting with different language-games and *choosing* to live according to one of them. Consider Wittgenstein's example of instructing a pupil on how understand the mark '→' on a signpost. To act in a way demanded by the sign requires following a rule which Wittgenstein claims we do 'blindly' (Wittgenstein 1953,

p. 85). Behaving in particular ways is the meaning of the mark. If we have someone who moves ← in the presence of the sign, they are *corrected*. They implicitly played a different language-game, according to which the arrow-head directs one to move in the opposite direction. Multiple language-games are necessary[10] for understanding in the first place.[11] Proliferation is built into the goal of understanding.[12]

Feyerabend also provides a second argument showing that pluralism is necessary for Wittgenstein's admission that language-games can *change* as a result of factual discoveries. (This assumption is also necessary for learning a language-game, since this requires teaching factual discoveries to a pupil). Wittgenstein claims:

> If anyone believes that certain concepts are absolutely the correct ones, and that having different ones would mean not realizing something that we realise – then let him imagine certain very general facts of nature to be different from what we are used to, and the formation of concepts different from the usual ones will become intelligible to him. (Wittgenstein 1953, p. XII)

In response, Feyerabend claims that

> I do not think he makes it clear that these 'very general facts', if they exist, could not make themselves noticed *except with the help of a conceptual system which takes them into account* and is therefore very different from whatever language is the commonly accepted ones. Such systems cannot be built up in a second. The appear in the form of non-testable, i.e., *philosophical* positions which conflict with the common views of the time . . . But their use, which is the first step towards a reform in language is excluded by Wittgenstein's demand that 'philosophy may in no way interfere with the actual use of language'. (Feyerabend 1969a, p. 159 fn. 29)

In other words, the process of *reforming* a language, one that is admitted as a possibility by Wittgenstein and his successors, requires pluralism and the violation of quietism.

---

[10] Josh Mozersky has pointed out to me that it is logically possible for people's behaviour to coincidentally always line up with the expectations of the language-game such that contrast is not, strictly speaking, necessary for understanding. This strikes me as correct, but given its empirical plausibility, I think this is a minor enough bullet for Feyerabend to chew on.

[11] Fascinatingly enough, this coincides with Feyerabend's endorsement of Mill's views that proliferation is necessary for understanding one's own views (see Feyerabend 1981c, 139).

[12] A persistent critic may point out that we may also ask 'why should we understand?' rather than act arbitrarily. While I have no definite response here, it seems like this critic already must have some understanding to launch that criticism, entailing that a criticism of Feyerabend's meta-philosophy requires the goals Feyerabend's meta-philosophy sets out to attain.

6.3 Feyerabend's Analysis of the Mind–Body Problem

Consider an alternative to materialism, dualism, wherein 'believing', 'hoping' and so forth are mental states or aspects of phenomenal consciousness. This has epistemic implications; we can immediately know, by consulting our own experience, whether or not we are believing. To say that 'Sally believes that she will win the pennant' refers to Sally having the appropriate mental state and *not* 'Sally is having a late slow wave in her right posterior.' Mental states are fundamentally distinct from brain states.

For Feyerabend, this could be viewed a perfectly fine *theory*. Putnam's argument for multiple realisability where mental states are underdetermined by brain states (Putnam 1967) could be viewed as predicting that there are no well-defined neurological correlates for mental states. Tests can be constructed to discern whether this is true (see Shapiro 2008). However, the mind–body problem is often not thought of where its 'solution' is a scientific theory. Rather, the defences of dualism Feyerabend rejects claim that dualism is viable because it is consistent with our current usage of concepts (i.e. it 'does justice' to our understandings of terms like 'belief') and preserves intuitions about bats, zombies, and Mary.

These defences suggest that the purpose of a philosophical theory must be consistent with how we already use particular concepts. Summarising Norman Malcolm's view, Feyerabend writes '[Malcolm's] argument against 'There are no material things' and his reason for calling this statement the expression of a 'philosophical paradox' would be as follows: the statement does not commit any mistake of fact. It is erroneous because it uses 'improper language'' (Feyerabend 1969a, p. 147; see also his 1963a, p. 162). For Feyerabend, this only shows that dualism is true within a particular language-game. Since this language-game provides an *ontology*, Malcom's implied methodological dictum demands that new theories should be consistent with old theories implicit in current language-games. But if we proliferate new uses of language, and demand that our intuitions be adjusted since they are wrong according to a new theory, then these appeals lose their force. Rejecting this new theory enforces an arbitrary conservatism where we prioritise our *current* intuitions and uses of language over future uses of language. As Feyerabend puts it,

> It is evident that this argument is incomplete. An incompatibility between the materialistic language and the rules implicit in some other idiom will criticise the former only if the latter can be shown to possess certain advantages. Nor is it sufficient to point out that the idiom on which the

comparison is based is in common use. This is an irrelevant historical accident. Is it really believed that a vigorous propaganda campaign which makes everyone speak materialese will turn materialism into a correct doctrine? (ibid).

This forbids the development of new theories (which would modify our language and intuitions as the Copernican hypothesis did), for no reason other than one happens to already be used.

As such, the mind–body problem has an easy answer: we consult a language-game and discern its ontology. However, we are also admonished to invent new language-games that will redefine the ontology with different answers. Each language-game will have a specified ontology, defined in whatever way with regard to the existence of mental states or their relationship(s) to material states, and evaluated as a scientific theory.[13] At an early, Popperian stage of his thinking, Feyerabend evaluates theories via their *testability*; theories must have as much empirical content as possible:

> The content of a statement increases with the number of statements which could count as refuting evidence. If the statement does not exclude anything, if it is valid in all possible worlds, then it is incapable of selecting situations of the real world in which we live, and is therefore dead weight. Languages containing such dead weight ought to be rebuilt to become more effective means of communication. (Feyerabend 1969a, p. 153)

More acutely, Feyerabend implicitly accepts Popper's view that for a theory to be *new*, it must have *different* empirical content (i.e. makes distinct predictions). As such, there are many alternative solutions to the mind–body problem, which can be compared on factual grounds exclusively.

6.4 Feyerabend on Materialism

Feyerabend's meta-philosophy repudiates views relying on linguistic usage. But there are purportedly non-linguistic and non-empirical arguments against materialism, which Feyerabend also addresses: the 'argument from observation' and the 'argument from acquaintance'. In this section, I recapitulate Feyerabend's arguments against these points. Before beginning, though, it should be noted that Feyerabend's defence of materialism

---

[13] This solution comes eerily close to Carnap's in 'Empiricism, Semantics and Ontology' (Carnap 1950).

does not commit him to materialism. It merely shows that these arguments do not foreclose the possibility that materialism could be true.[14]

The argument from observation runs as follows: 'That a thought cannot be a material process is, so it is believed, established by observation. It is by observation that we discover the difference between the one and the other and refute materialism' (Feyerabend 1963a, p. 165). By 'observation', Feyerabend means 'introspection'. He admits that introspection 'indicate-[s], in a most decisive fashion, that my present thought of Aldebaran is not localised whereas Aldebaran is localised; that this thought has no colour where Aldebaran has a very definite colour; [etc.]' (ibid). The response is straightforward and will be familiar to Feyerabend aficionados: appearance should not be conflated with reality and our theories should explain reality and their relationship to appearances.[15] The way a thought appears may be different from its physical manifestations in the same way a 'heard sound is very different from its mechanical manifestations' (ibid). Identification is an empirical hypothesis and materialists can hypothesise about what the correlates may be.[16]

The argument from acquaintance goes back to Russell, at least, who claims that 'we have [knowledge by] *acquaintance* with anything of which we are directly aware, without the intermediary of any process of inference or any knowledge of truths' (Russell 1912, p. 78). The argument runs like this:

> Mental processes are things with which we are directly acquainted. Unlike physical objects whose structure must be unveiled by experimental research … they can be known completely and with certainty. Essence and appearance coincide here, and we are therefore entitled to take what they seem to be as a direct indication of what they are. (Feyerabend 1963a, pp. 166–167)

Materialism denies this since it presumes that introspection can be mistaken and this can be shown by studying material processes. I can experience pain but would be wrong to say 'I am in pain' on the materialist view since 'pain' refers to some material process, which may or may not line up with particular mental states. The certainty of 'I am

---

[14] This is consistent with and extends my conception of Feyerabend's critical philosophy as a series of arguments against 'rationalism' or the view that we can, in principle, exclude views from possibly being true (Shaw 2020).

[15] 'I presuppose a minimum of reason in my readers; I assume they are realists' (Feyerabend 1963a, p. 166).

[16] To be clear, Feyerabend does not have any (inherent) qualm with introspection as a methodology insofar as, with the help of an identification hypothesis, it is construed as a direct observation of the brain. He even suggests that this may be more likely that observations of dead tissues (ibid.; note that this was before reliable in vivo methods).

in pain' is 'a logical certainty: there is no possibility whatever of criticizing the statement' (p. 167). Feyerabend claims the source of this certainty is their lack of content.

> Statements about physical objects possess a very rich content ... Thus, the statement 'There is a table in front of me' lead to predictions concerning my tactical sensations; the behaviour of other material objects [etc.] ... Failure of any one of these predictions may force me to withdraw the statement. This is not the case with statements concerning thoughts ... No prediction, no retrodiction can be inferred from them. (pp. 167–168)

Two problems arise from this. One, which has already been mentioned, is that propositions with no empirical content do not express anything at all. 'I am in pain', interpreted as being known directly by acquaintance, does not allow us to infer anything:

> If all these elements are removed, then what do I mean by the new statement ['I am in pain' interpreted as known by acquaintance] resulting from this semantic canvas cleaning? I may utter it on the occasion of pain (in the normal sense); I may also utter it in a dream with no pain present; I may use it metaphorically, connecting it with a thought (in the usual sense) concerning the number two; or I may have been taught ... to utter it when I have pleasant feelings and therefore utter it on these occasions. (Feyerabend 1965a, pp. 195–196)

As such, this language should be abandoned on *methodological* grounds. However, Feyerabend offers a second, stronger argument that using the knowledge of acquaintance to establish that materialism is false is circular:

> If we possess knowledge by acquaintance with respect to mental states of affairs ... then this is the *result* of the low content of the statements used for expressing this knowledge. Had we enriched the notions employed in these statements in a materialistic ... fashion *as we might well have done*, then we would no longer be able to say that we know mental processes by acquaintance. (Feyerabend 1963a, p. 170)

The argument that we have knowledge by acquaintance presupposes the dualistic metaphysics it provides an argument for. Neither of these arguments can establish the validity of dualism.

6.5 Feyerabend's Change of Mind

When 'Linguistic Arguments and Scientific Method' was reprinted, Feyerabend added an additional footnote claiming that he had changed his mind:

> It must be admitted that the overthrow of an entire world view. ... can be
> stopped by the *decision* to make commonsense (and the views of man it
> contains) an essential part of any form of knowledge. Such as a decision was
> made by Aristotle. (Feyerabend 1969/1981, p. 146 fn. 1)

This is in response to his earlier claim that 'Dogmatic philosophies therefore do not advance our knowledge, all they do is describe ordinary and
well-known things in an extraordinary and not so well-known fashion' (p.
154). In the original paper, Feyerabend assumes that theoretical differences
are differences in empirical content. Feyerabend's revised view is that
theories are distinct if they promote different kinds of lives. This is also
mentioned in another added footnote in the same collection:

> The choice of an idiom for the description of mental events cannot be
> decided by considerations of testability and 'cognitive content' alone. It may
> well be that a materialistic language ... is richer in cognitive content than
> commonsense ... But it will be poorer in other respects. For example, it will
> lack the association which now connect mental events with emotions, our
> relations to others, and which are the basis of the arts and the humanities.
> We therefore have to make a choice: do we want scientific efficiency, or do
> we want a rich human life of the kind now known to us and described by our
> artists? The choice concerns the *quality of our lives* – it is a *moral* choice.
> (Feyerabend 1963/1981, p. 163 fn. 1)

This is spelled out in more detail in *Conquest of Abundance* with what he
calls 'Aristotle's Principle'.[17]

In his early career, Feyerabend claims that ethics forms the basis of
epistemology (Feyerabend 1961/1981, p. 55; 1965a, p. 219 fn. 5). This
proclamation is given more substance in his later writings where he out-
lines 'Aristotle's Principle' (see Brown 2016). Aristotle's principle states
that what is 'real is what plays a central role in the kind of life we identify
with' (Feyerabend 1999, 201). Aristotle's principle accepts that there is 'a
deeper lying stratum that responds positively to many different endeavors'
(p. 203) called 'Being'. Being provides 'hints, or analogies', as opposed to
'elementary building stones', for constructing theories (p. 204). Moreover,
'non-scientific notions, too, receive a response from [Being]' (p. 195).
Myths, theology, aesthetics, and so forth can influence our interactions
with Being. Along with science, these jointly create 'manifest worlds they
can expand, explore, and survive in' (p. 204). Aristotle's principle allows for

---

[17] Similarly, Rorty claims that 'we need many different descriptions of ourselves – some for some
purposes and others for others, some for predicting and controlling ourselves and others for deciding
what to do, what meaning our lives shall have' (Rorty 1982, p. 345).

different ways of living to incorporate empirical findings in various ways, though not *all* ways will be able to incorporate them in the desired way; Being resists our attempts to live in it in a specific way (Tambolo 2014). However, it is 'pliable' enough to allow for many different theories leading to many different kinds of life. This choice entails that our acceptance of a theory of reality is partially dependent on values.

Aristotle's principle can be seen more vividly in Feyerabend's analysis of the Galileo affair. The (second) Inquisition in 1632/3 made two evaluations; one concerning the empirical content of Galileo's astronomy and the other concerning its social/ethical ramifications. From an empiricist point of view, Feyerabend claims that the Church's evaluation of Galileo's astronomy as 'unscientific' was correct (Feyerabend 1975/1993, p. 128). On the second point, the Church claimed that Galileo was formally heretical since it contradicted Holy Scripture. On this, Feyerabend writes that

> we notice that it deals with a subject that is gaining increasing importance in our own times – the quality of human existence. Heresy, defined in a wide sense, meant a deviation from actions, attitudes and ideas that guarantee a well-rounded and sanctified life. Such a deviation might be, and occasionally was, encouraged by scientific research. Hence, it became necessary to examine the heretical implications of scientific developments. Two ideas are contained in this attitude. First, it is assumed that the quality of life can be defined independently of science, that it may clash with demands which scientists regard as natural ingredients of their activity, and that science must be changed accordingly. Secondly, it is assumed that Holy Scripture as interpreted by the Holy Roman Church adumbrates a correct account of a well-rounded and sanctified life. (p. 130)

Feyerabend contends that the Church was 'on the right track' with the first assumption, and partially rejects the second, and claims that while conceding that the 'Bible is vastly richer in lessons for humanity than anything that might ever come out of the sciences' (p. 131). The Church had the right 'intention' to 'protect people from being corrupted by a narrow ideology that might work in restricted domains but was incapable of sustaining a harmonious life' (p. 133). Aristotle's principle is on display here. Scientific theories can be rejected, or at least rejected as 'true', because they do not aid people in living meaningful lives. The Church was wrong, according to Feyerabend, for its undemocratic means of determining value; but the attempt to preserve a quality of life was a justifiable endeavour.

In this case, 'science' *conflicts* with theology. But, Feyerabend is playing fast and loose with what counts as 'science'. He is calling Galileo's astronomy 'science' because his readers have some pre-theoretical demarcation criteria. One may draw the distinction methodologically; science is what aims to predict future states of affairs. If this view is held, then 'non-science' and 'science' can compete, like how Nostradamus's prophecies compete with the sciences, forcing us to choose between them. Choosing the theory without strong support, without further ado, would be an instance of 'wishful thinking' (see Anderson 1995). But, there are many examples where competition is unnecessary. Consider Duhem's intermixture of philosophy of science and theology. For Duhem, scientific theories aim at a 'natural classification' of experimental laws rather than a metaphysical understanding of nature. This understanding is provided by theology, which has important ethical implications (see Martin 1991). Similarly, consider Galileo's view that the domain of science is empirical reality and the domain of religion is salvation. Principles on how to live are provided by *both* theories jointly. Predictive inequivalence is unimportant here, since understanding salvation and divine demands does not, for Galileo, require empirical tests.[18] The same is true of the Hindu belief that cows are composed of Gods, which leads to particular norms concerning diet and the treatment of cows. Hindus, most of the time, at least, do not try to argue against biological or zoological theses about cow anatomy or behaviour (see O'Flaherty 1979). The treatment of cows follows from their symbolic place in *The Vedas* and the belief that they should be respected for their production of useful goods. Here, science as an instrument of prediction is methodologically segregated from other domains of discourse.

But, there is another dimension to Aristotle's principle. Consider Carnap's reframing of metaphysics as akin to poetry:

> Thus, for instance, a Metaphysical system of Monism may be an expression of an even and harmonious mode of life, a Dualistic system may be an expression of the emotional state of someone who takes life as an eternal struggle; an ethical system of rigourism may be expressive of a strong sense of duty or perhaps of a desire to rule severely. Realism is often a symptom of the type of constitution called by psychologists extraverted, which is characterized by easily forming connections with men and things; Idealism, of an opposite constitution, the so-called introverted type, which has a tendency to with draw from the unfriendly world and to live within its own thoughts and fancies. (Carnap 1935, pp. 29–30)

---

[18] For a general discussion of conceptions of the relationship between religion and science, see Finocchiaro (2001).

Similarly, Hempel claims that the goal of logical empiricism is to eliminate meaningless propositions regardless of 'however rich some of them may be in non-cognitive import by virtue of their emotive appeal or moral inspiration they offer' (Hempel 1950, p. 41). For Feyerabend, even if there is no cognitive difference between scientific and 'non-scientific' theories *and* there is no segregation of their domains, the *rhetorical* or *symbolic* quality of theories can still be of moral significance. Eric Martin puts this point well:

> Feyerabend takes seriously such psychological and emotive contours of philosophies like materialism. If the adoption of materialism was tried and found 'depressing', that could count as a mark against its acceptance.[19] Such consequences are not incidental features of those philosophies, revealing the wishful thinking of hapless philosophers and critics of science. Rather, they are features to be seriously evaluated as a part of the democratic critique of science he advocates. (Martin 2016, p. 132)

A similar statement of this view, though about moral statements rather than scientific theories, can be found in Nietzsche:

> To this extent moral judgment is never to be taken literally as such it never contains anything but nonsense. But as *semiotics* it remains of incalculable value: it reveals, to the informed man at least, the most precious realities of cultures and inner worlds which did not *know* enough to 'understand' themselves. (Nietzsche 1968, p. 55)

Perhaps 'God exists' can be translated into an empirically adequate scientific theory *salva veritate*, but this translation must *also* preserve the quality of the statement that gives meaning to the people who consider it. Would this God inspire the same artwork? Would the same traditions or festivals follow from the scientific God?[20] Empirical equivalence cannot answer these questions. For Feyerabend, these rhetorical differences can be of *paramount* importance; they determine what makes life meaningful and provide existential guidance.

Notice what consequences this has for Feyerabend's naturalism. In one sense, Aristotle's principle follows from Feyerabend's naturalism. If theories are parts of the natural world, then being morally considerate of the psycho-sociological consequences of their acceptance or usage makes sense.

---

[19] For a related discussion, see Disch's (1996) discussion of Hannah Arendt's views on the limitations of abstractly formulated theories and the advantages of storytelling. This view has Feyerabendian affinities given Feyerabend's repeated likening of scientific theories to fairy-tales.

[20] This reveals implications of Aristotle's principle for the literature on verbal disputes where the cognitive inter-translation of two theories shows they involve 'logically irrelevant changes of typography' (Quine 1951, 210; see Hirsh 2016). Even if this is correct, it is insufficient for the acceptance of languages that are epistemically identical but otherwise distinct.

But, we should also notice that Feyerabend's naturalism is broader than many contemporary formulations. For some, methodological naturalism is the view that what counts as real is what we are committed to according to our best scientific theories (see Papineau 2014). But, for Feyerabend, this is too narrow a source for our ontological commitments. We can also consult 'non-scientific' theories for ontological commitments. Moreover, the aforementioned pliability allows for us to remain pluralists and naturalists simultaneously. Naturalism does not impose a particular ontology, especially in the case of societies with diverse needs and backgrounds.

But AP is not supporting a foundationalist enterprise where the acceptance of scientific theories is parasitic on the acceptance of ethical theories. Rather, it admonishes a *dialogue* between ethical and scientific claims. If the former reading were true, it would conflict with Feyerabend's claim that ethical theories can be abandoned as a result of scientific discoveries (Feyerabend 1970e, p. 127; 1975a, p. 189). Rather, 'Ethics (in the general sense of a discipline that guides our choices between forms of life) affects ontology. It already affected it . . . but surreptitiously, and without debate' (Feyerabend 1999, p. 247). The parenthetical remark suggests that we shouldn't think of ethical *theories* as guiding theory-choice but *disciplines*, within which there are debates. Aristotle's principle does not force us to exclusively pursue theories with firm ethical foundations; it merely admonishes us to discuss ethical issues in determining what counts as real.

We are now in a position to understand Feyerabend's implicit use of Aristotle's principle in his later criticisms of materialism and his changed assessment of the mind–body problem. As Martin documents, Feyerabend became interested in how religious and mystical worldviews provided alternative frameworks that supersede materialism in their moral qualities:

> Materialism's shortcomings and potential for harm are both symptomatic of laws according to what might be called its *existential* criteria: the way that a philosophy contributes to one's most general understanding of the world, one's place in it, and one's actions in that world. (Martin 2016, p. 132)

Additionally, Feyerabend does not merely think that materialism has this *capacity*, which religion and mysticism *may* overcome, he declares this to be the case: 'For Feyerabend, those with religious proclivities or temperaments, open to the world's mysteriousness or responding to the world with a sense of reverence, might well lead richer or more humane lives' (131). This being said, I agree with Martin that 'Feyerabend's own writing on the topic is not necessarily the most fruitful or nuanced to be found' (134). Feyerabend's accusations are grounded in fairly crude, inarticulate forms of materialism and, similarly, relatively superficial assessment of religiosity and mystical traditions. While Feyerabend was well informed on these subjects, his texts are too sketchy to take his anti-materialism at face value. However, Feyerabend offers a deeper level of analysis in two respects. First, his earlier defences of materialism only suggest that materialism *may be* true; it does not tell us what to accept, whereas his later work provides a proposed assessment of materialism. Second, it provides us with a new existential means of evaluating theories that goes beyond methodological dicta. Even if Feyerabend's texts are too bare to provide us with solid answers, they provide a path forward for to resolve the mind–body problem.

6.6 Concluding Remarks

As a final note, it is worth drawing an interesting parallel between the development in Feyerabend's thought and the development of pragmatism. For some, William James' dictum that we should judge what is true by its practical consequences is a naïve articulation of more plausible formulations, such as those of Peirce or Ramsey, which claim that truth as a correspondence between theories and the world remains an ineliminable part of pragmatism (Misak 2018). If this is right, then Feyerabend's 'development' appears to be backwards if we think of 'Aristotle's principle' as analogous to James's maxim and his earlier naturalist empiricism as similar to the views of other pragmatists. But such a comparative assessment and evaluation in either case is beyond the current scope of this paper. Given that the links between Feyerabend and the pragmatists have yet to be discussed in serious depth, this topic seems to be a fruitful terrain for future discussion.

As every historian knows, a closer look at the texts of many discussed figures reveals that their views are more nuanced and interesting than the slogan versions that are passed down. It is my belief that Feyerabend's treatment of the mind–body problem is an example of this phenomenon. Given that contemporary philosophers of mind scarcely engage with Feyerabend's views, it makes sense to suggest that such engagements will lead be fruitful. Of course, such engagements have not been directly attempted in this chapter and future attention is needed to see whether this hunch is correct. But still, given the gravity of the consequences of Feyerabend's position, such engagements seem worthwhile for future considerations.[21]

---

[21] Acknowledgements: I greatly appreciate the recommendations of Peter Verveniotis, Nora Hangel and Josh Mozersky which helped direct the research for this chapter.

CHAPTER 7

Feyerabend's Re-evaluation of Scientific Practice
Quantum Mechanics, Realism and Niels Bohr

Daniel Kuby

7.1 Introduction

In this chapter, I offer a specific interpretation of how Feyerabend came from a Popperian critique of the Copenhagen interpretation to a detailed re-evaluation of Niels Bohr's idea of complementarity. Engaging with this chapter of Feyerabend's intellectual *Werdegang* is not only an interesting exercise in Feyerabendian exegesis; an explanation of this change of mind in a very narrow domain – or so it seems – gives the backdrop for Feyerabend's thoroughgoing turn from methodological monism to methodological pluralism, for which he would became known to a wider audience with his publication of *Against Method* (Feyerabend 1975a).[1]

In his early philosophy and until the mid-1960s, even though he used historical case studies, Feyerabend positioned himself decidedly against the wave in philosophy of science that would eventually be labelled as its 'historical turn'.[2] This is ironic in light of the fact that Feyerabend is remembered to this day as a proponent of the turn. Though his later adherence to the turn is not disputed, it is also recognised that his previous philosophical stance had a normative urgency towards the right

---

[1] The distinction between an 'early' and a 'later' Feyerabend has been canonised by Preston (1997a), according to whom Feyerabend's 'work can be (roughly) divided into two phases, the first stretching from the early 1950s until about 1970, the second from 1970 onwards' (Preston 1997a, p. 7). Oberheim (2006, pp. 15–16) has sensibly objected that many of Feyerabend's post-1970 ideas can (quite literally) be traced back to earlier writings throughout the 1960s. I agree with Oberheim, but I contend that there is a wall we hit in the back-dating game, around 1964–1965, such that famous articles usually classified as pertaining to the historical turn (like Feyerabend 1962a; Feyerabend 1965a) cannot and should not be assimilated to it. I also agree with Oberheim that the continuity in Feyerabend's philosophical work is much better located at the level of his metaphilosophical commitments, though I give a rather different interpretation of what these are (cf. next footnote).

[2] See e.g. the letters from Feyerabend to Kuhn on a draft of *Structure of Scientific Revolutions*, (edited in Hoyningen-Huene 2002; Hoyningen-Huene 2006) for some evidence. Hoyningen-Huene (2002, pp. 68–72) repudiates Feyerabend's criticism of Kuhn as a misunderstanding of Kuhn's position. My (2019a) agrees with Hoyningen-Huene, but gives an explanation for this misunderstanding based on Feyerabend's metaphilosophical background as developed in my (2019b).

methodology to be used in the sciences that his later philosophy would lack. Indeed, I propose to recognise this normative stance as a kind of *philosophical prescriptivism*, according to which philosophy of science qua general methodology has standing to make prescriptive claims vis-à-vis the sciences. This view grew particularly strong in the early 1960s, in that only general methodology has standing to set up methodological rules. Still a methodological monist, Feyerabend defended a consistent set of methodological rules, most importantly the demand to interpret our best scientific theories realistically, as means to realise the core scientific value of testability. I reconstruct Feyerabend's methodological argument for realism as follows:

P1:   Theory-testing is a constitutive task of science. (Principle of Testability).
P2:   Interpreting scientific theories realistically is more likely to maximize their testability than alternative interpretations.
P3:   If scientists want to test scientific theories, they should interpret scientific theories realistically.

Conclusion: Scientists should interpret scientific theories realistically. (Realism)

The justification of testability as a core scientific value, however, was based in a purely axiological decision concerning the aims of science (see Feyerabend 1961/1981 for a statement of this view and Kuby 2019b for discussion).

The first observation we can make is that Feyerabend's adherence to the 'historical turn' coincides with an abandonment, indeed a rejection of this *philosophical* prescriptivism (normative claims could be raised only contextually to a specific research situation). A more specific question about Feyerabend's adherence to the turn can, then, be asked in terms of how Feyerabend came to abandon his prescriptivism and which factors made him abandon it. One avenue of research is to relate the dynamics of Feyerabend's philosophical views to a broader context, by noting the political and social turmoil that coincides with his changing views – most notably, the effects of desegregation around US-American universities (starting in 1954) and the Free Speech Movement at UC Berkeley (starting in 1964), where Feyerabend had been a tenured Professor since 1959. Surely, no explanation of Feyerabend can be complete without putting this picture at the centre stage. In this chapter, however, I will put forward an explanation of Feyerabend's philosophical journey in a complementary fashion, in an attempt to resist the narrative of an 'anarchic overturn' between the early and later Feyerabend (as has been given in Preston 1997a). In fact, I will offer a very standard explanation in terms of a simple model of change of belief through evidence.

My main claim in this chapter is that the evidence he was exposed to came through his engagement in (the history of) quantum mechanics, in particular with a re-evaluation of Neils Bohr's contribution to it. I contend that Feyerabend's prescriptivism was first confronted with a serious problem in the specific context of his methodological arguments for realism vis-à-vis justified scientific practice in quantum mechanics. A crucial feature of Feyerabend's methodological argument for a realistic interpretation of scientific theories is its generality. The argument is universal in scope, such that the methodological demand obtains 'for all scientific theories'. It poses no conditions on its application on the specifics of a theory, in part, because the argument applies to scientific theories as reconstructed in the statement view, which completely abstracts from the specifics of any given scientific theory. It was the universal scope of the argument that was slowly but steadily put into question. Throughout the 1960s, Feyerabend recognised specific instances of arguably scientific theories for himself, in which differing demands were legitimate because they 'made sense scientifically', putting a dent into Feyerabend's top-down methodological argument scheme: for a specific research situation, we arrive at contrasting demands whether we look at it from a general–methodological or from a contextual-scientific point of view. Such was the situation of Bohr's interpretation of quantum mechanics. This is what I want to call *Feyerabend's dilemma*:

1. According to Feyerabend's philosophical prescriptivism, the compelling reasons for a specific scientific behaviour are axiological.
2. There's a specific class of scientific behaviour that Feyerabend finds compelling, for reasons that are independent from axiological considerations.[3]

As long as the methodological conducts derived from (1) and (2) are compatible, no problem arises. It might not even be possible to conceptually separate (2) from (1), that is, to be forced to recognise that the

---

[3] I claim that Feyerabend's conception of philosophy of science, even at its prescriptivist peak, is not free from a task inherited from previous philosophies of science: whatever scientific method is proposed, it has to account for the shared canon of modern science. Here Feyerabend gives probative value to the canon of past scientific achievements in the form a shared set of exemplary contributors to science (what Laudan (1989, p. 213) calls 'the Tradition'), independently from axiological considerations. As Laudan (1989) has pointed out, one of the main tasks of new positions defended in philosophy of science has been to reinterpret the Tradition to fit their view on methodology. Even if the Tradition varies in scope, prototypical scientists like Galileo, Newton and Einstein are there: [A]ny proposals about the aims [and methods] of science must allow for the retention as scientific of much of the exemplary work currently and properly regarded as such. A suggested aim [and method] for science which entailed, for instance, that nothing in Newton's *Principia* was really scientific after all would represent such a distortion of scientific practice that it would be wholly uncompelling' (Laudan 1990, p. 47).

behaviour in (2) is not chiefly dependent on axiological decisions. The problem arises if the methodological behaviours derived from (1) and (2) are incompatible.

This contrast became more and more strident, until Feyerabend was forced to give up the universality of the methodological argument, which initiated a cascade of consequences extending to the very core of his conception of what philosophy of science is about. He had discovered and came to acknowledge the existence of a scientifically justified, theory-specific notion of ontological interpretation, which stood in contrast to a theory-independent, axiologically justified notion of ontological interpretation. I contend that the source responsible for bringing Feyerabend face to face with this evidence was his physical and philosophical interest in quantum mechanics.[4]

I will proceed as follows: Section 2 gives the set-up of the early Feyerabend as a philosopher chiefly preoccupied with quantum physics and its philosophy, who attacked the 'Copenhagen interpretation' as an instrumentalist interpretation of quantum mechanics along Popperian lines. I also note that we can pinpoint at first a timid change of mind with regard to Niels Bohr to 1957. In Section 3, I argue that, motivated by this change of mind, Feyerabend's early goal was to disentangle, respectively, the philosophical and the empirical support for the initial introduction of the principle of complementarity in quantum mechanics. I then distinguish between the vindication of complementarity in quantum mechanics, which Feyerabend came to appreciate on empirical grounds, and complementarity as a general requirement for future microphysical theories, which Feyerabend still rejected on philosophical grounds. In Section 4, I introduce Feyerabend's framework for distinguishing syntax, semantics and ontology of scientific theories in order to understand his original interpretation of Bohr's complementarity as an ontological interpretation grounded in physical arguments. This construal leads to a first instantiation of 'Feyerabend's dilemma': with regard to the same theory, philosophical reasons compel us to demand a realistic interpretation, while physical reasons compel us to accept an instrumentalist interpretation. In Section 5, I show that Feyerabend reduced the problematic import of the dilemma by appealing to his theoretical pluralism. Indeed, the dilemma

---

[4] I want to stress that Feyerabend's change of mind cannot be explained by his exposure to this evidence alone. He had to be receptive to this evidence in the first place. This receptiveness is rooted in specific characteristics of his overall metaphilosophical conception, in particular, the demand that methodological rules should be actually realisable, which had become almost ineffective in his philosophical prescriptivism and slowly regained importance. For a reconstruction of Feyerabend's metaphilosophy as 'decision-based epistemology', see Kuby (2019b).

could be used for an argument for the principle of proliferation, demanding realistically interpreted alternative theories in microphysics. In Section 6, I argue that the dilemma was only circumvented, not resolved. I argue that Feyerabend maneuvered himself into too strong an argument for realism in microphysics by showing that the general inference from testability to realism is not viable any more (while leaving the inference to proliferation intact). Section 7 concludes by arguing that Feyerabend came to this realisation, too. More than that, he viewed this failure as a refutation of his own philosophically motivated, theory-independent notion of realism, leading him to embrace the methodological constraints put in place by scientific theories and the local and fallible status of methodological rules, to which Bohr's contribution in quantum mechanics he now considered a testament.

7.2   Feyerabend and Quantum Mechanics

Feyerabend is often thought of as a philosopher working in general philosophy of science – and, since his (1970e), openly advocating its demise. It might therefore come as a surprise that Feyerabend started out as a philosopher of quantum mechanics. His early scholarly production deals overwhelmingly with problems in microphysics (Feyerabend 1954a; 1956; 1957a; 1957b; 1957c; 1958a; 1958b; 1958c; 1960a; 1960b; 1960c; 1961; 1962b; 1963b). In earlier as well as later papers quantum mechanics continued to surface as historical casuistry. Feyerabend got to work on quantum mechanics from the late 1940s as a trained physicist with an interest in philosophy of physics. His earliest extant paper (Feyerabend 1948/2016), written as an undergraduate student, deals with the concept of intelligibility in microphysics (cf. Kuby 2016). We do not know the exact topic of his attempted dissertation thesis, which was to deal with classical electrodynamics – but we know that he abandoned it to work on the philosophical problem of basic statements (Feyerabend 1951).

We have only scarce evidence on how Feyerabend came to concentrate on quantum mechanics. It coincides temporally with his stay abroad at the London School of Economics in 1952 with Karl Popper and took off from there (cf. Feyerabend 1995, p. 92). Feyerabend came to work on a large number of topics: indeterminism in the microphysical domain; the limits of the von Neumann no-go theorem; the quantum theory of measurement; quantum mechanical formalisms, in particular quantum logic; ontological interpretations of quantum mechanics; and alternative theories to quantum mechanics. Feyerabend's philosophical allegiance to Popper consolidated around that time. On his return to Vienna, his first research project in 1954 included an analysis of 'the role of the ergodic hypothesis within classical statistics' as part of the larger topic 'The function of hypotheses in science', on which Feyerabend remarked in a letter to Popper: 'the title already mirrors your influence' (Feyerabend to Popper, 12 March 1954, in Feyerabend 2020, p. 117).[5] At first, very likely due to this intellectual bond, Feyerabend came to adopt Popper's specific criticism in the philosophy of quantum mechanics, chastising its main proponents as giving in to an unwanted and unwarranted positivist position (cf. Feyerabend 1954a/2015, p. 34; Feyerabend 1954b/2015, p. 12), which allegedly had been built on the scientific consensus at the Fifth Solvay Conference in 1927 and was ascribed to the Copenhagen–Göttingen school of Niels Bohr, Werner Heisenberg, Max Born and Wolfgang Pauli. Feyerabend repeatedly invoked Popper's sweeping picture of a capitulation of physics to vicious philosophy in his early papers, too:

> Today the view of physical science founded by Osiander, Cardinal Bellarmino, and Bishop Berkeley, has won the battle without another shot being fired. Without any further debate over the philosophical issue, without producing any new argument, the instrumentalist view (as I shall call it) has become an accepted dogma. It may well now be called the 'official view' of physical theory since it is accepted by most of our leading theorists of physics (although neither by Einstein nor by Schrödinger). And it has become part of the current teaching of physics. (Popper 1956, p. 360)[6]

But then something changed. In his (1958a), for the first time Feyerabend timidly used a footnote to exonerate the founder of the Copenhagen school from the charge of deceivingly stating the Copenhagen interpretation as a necessary consequence of the formalism of quantum mechanics.[7] In private correspondence, we can predate a change of mind about Bohr already to an earlier time. In a short post scriptum to a letter to Popper, Feyerabend notes that 'there is much more in the Copenhagen-interpretation (as it has been discussed by *Bohr*, *not* by the Bohrians) than I thought some time ago when

---

[5] Preston (1997a) highlights the Popperian heritage of many themes in Feyerabend's early philosophy. Farrell (2000, 275) claims that even the later Feyerabend was 'in many respects, a die-hard pluralistic Popperian'. For a more nuanced view of the institutional and philosophical relationship between Feyerabend and Popper, see Collodel (2016).

[6] Popper's capitulation picture is important because it licensed the use of the Copernican Revolution as a foil to discuss the interpretation of quantum mechanics, and one can track Feyerabend's position by the way in which he handled Popper's thesis both regarding the Copenhagen interpretation and the Copernican Revolution.

[7] A charge levelled in that context against Von Neumann's (1932/1955) presentation of the theory; Feyerabend (1958a, p. 346 fn. 1) exonerates Bohr in one succinct remark without further comments: 'It ought to be mentioned that Bohr himself did not commit this mistake.'

I did not know it well enough' (Feyerabend to Popper, 21 July 1957, in Feyerabend 2020, p. 267).

What happened in 1957? Feyerabend's engagement with the original literature of the 'first quantum revolution' coincides with the Ninth Symposium of the Colston Research Society, hosted by the University of Bristol (April 1st–4th), where Feyerabend held his first appointment as lecturer in philosophy. The conference was seminal in challenging the scientific orthodoxy after World War II and helped create a climate in which philosophers of science considered foundational issues to be open questions again, creating a platform for challenges to (and defences of) scientific orthodoxy – though these issues would be accepted back into physics only with Bell (1964; see Kožnjak 2017).

Though the evidence is sparse, discussions during the conference also alerted Feyerabend to the fact that his knowledge of Bohr's own views and arguments were deficient. In particular, he was made aware that his contribution to the conference, on the topic of quantum measurement theory, was not a counterpoint to Bohr's view, as Feyerabend framed it, but much along lines that Bohr had previously indicated.[8] This gave Feyerabend pause – not in his philosophical struggle against positivism and subjectivism in quantum theory, but in associating *Bohr*'s position with positivism and subjectivism. Given Bohr's key role in the development of quantum theory, Feyerabend developed a genuine interest in his ideas, which would have deep repercussions at the very core of Feyerabend's metaphilosophy. But first, this new perspective on Bohr's work ignited a series of detailed examinations of Bohr's contribution to quantum mechanics, recognising his unique perspective (Feyerabend 1958c; 1961; 1962b; 1968a; 1969b).

7.3   The Role of Physical Argument: Feyerabend Re-evaluates Bohr

Feyerabend's motivation to learn about the original development of quantum mechanics was greatly enhanced by his participation in the Colston Symposium. Access to the original development of quantum mechanics meant access to the dynamics of scientific reasoning behind its establishment: *Did complementarity earn its place in microphysics? If so, how?* Nothing less was the motivation of Feyerabend's interest in the early history of quantum mechanics. Popper had taught him, mostly on general methodological grounds, that complementarity had not earned its place.

---

[8] An account of this incident and of Feyerabend's contribution to the rise of the quantum measurement problem is in preparation.

Feyerabend first followed Popper, but then came in contact with historical protagonists, the original literature and he started to think differently. We can see this progressive awareness in an almost chronological ordering of his papers: In Feyerabend (1958c), Feyerabend wants

> to show that [Bohr's point of view] is consistent, that it has led to important results in physics and that it therefore cannot be easily dismissed. It will also turn out that this point of view is closely related to the position of positivism: the issue between the classical model of explanation and complementarity is essentially an instance of the age-old issue between positivism and realism. (Feyerabend 1958c, pp. 80–81)[9]

Initially, he recognised Bohr's notion of complementarity as a proposal for a new model of scientific explanation. This model diverges from the classical model of explanation in how it treats the two groups of experimental facts that firmly established the wave–particle duality of light. Two theories can completely explain each group of facts, yet they are mutually exclusive. While the classical model of explanation regards 'the existence of two non-exhaustive and complementary descriptions ... to be an historical accident, an unsatisfactory intermediate stage of scientific development' to be hopefully solved by the 'search for a new conceptual scheme', the new model accepts the duality and changes the very requirements of what a scientific explanation is. The classical model demands that 'such a new theory ... must be empirically adequate, i.e. it must contain the facts [about duality] as approximately valid under mutually exclusive conditions ... [a]nd it must be universal, i.e. it must be of a form which allows us to say what light *is* rather than what light appears to be under various conditions' (p. 78). In this sense, it is 'closely connected with the position of realism' (p. 79). Bohr instead does not regard duality as 'a deplorable consequence of the absence of a satisfactory theory, but a fundamental feature of the macroscopic level. For him the existence of this feature indicates that we have to revise ... the classical *ideal of explanation*' (p. 79). This new ideal of explanation, expressed in the principle of complementarity, 'does not consist in relating facts to a universal theory, but in their incorporation into a predictive scheme none of whose concepts is universally applicable' (pp. 87–88). It is therefore an abdication of realism in that it not only a) gives up universal applicability of quantum-mechanical concepts as a condition of explanation, but b) by replacing traditional theories with the notion of 'natural generalization of classical physics' following the correspondence rule, also of any future quantum theory (p. 90).

---

[9] Yet Bohr's work stands in contrast to other physicists of the 'Copenhagen school ... To them Popper's remark [about the capitulation of physics, see above] applies' (Feyerabend 1958c, p. 80).

Is this abdication justified? Feyerabend maintains that this new model of explanation is successful in the case of quantum mechanics and he gives a first rundown of how complementarity fits well with elementary quantum theory. In this sense, (a) can be said to be justified, though with important limitations. But Feyerabend argues vehemently against (b): the new model makes the classical ideal of explanation, which it tries to replace, neither impossible nor obsolete; more importantly, its application to the very possibility of future physics would lead to a complete 'stagnation' of physics (pp. 103–104).

Next, Feyerabend went into a detailed examination of the source literature to appreciate Bohr's interpretation not just as a philosophical preconception that happened to be physically successful, but as an outcome of scientific research, a point he argued at length in his papers 'Niels Bohr's interpretation of the quantum theory' (1961) and 'Problems of microphysics' (1962b) which incorporated and expanded his (1961), and reaffirmed much later in his long two-part paper 'On a recent critique of complementarity' (1968a; 1969b), prompted by Mario Bunge (1967), which Feyerabend deplored. It was in 'Problems of microphysics' that for the first time he put the (mostly qualitative) *physical reasoning* at the centre stage: Bohr's 'point of view can stand upon its own feet and does not need any support from philosophy' (Feyerabend 1962b, p. 292). He lays out the main aim of his paper as follows:

> I shall try to give a *purely physical explanation* of the main ideas behind the Copenhagen Interpretation. It will turn out that these ideas and the *physical arguments* leading up to them are much more plausible than the vague speculations which were later used to make them acceptable. (p. 195, emphasis added)

I want to draw attention to the emergence of the notion of 'physical arguments' as a crucial step in Feyerabend's re-evaluation of Bohr's contribution to quantum mechanics. To sustain the Copenhagen interpretation, says Feyerabend, 'much better arguments are available, arguments which are directly derived from physical practice' (p. 194). For him, scientific practice as seen through the dynamics of physical arguments can deliver reasons for understanding and evaluating scientific decisions. And, most important of all, the class of 'physical arguments' gives us an instantiation of step (2) in Feyerabend's dilemma, that is, a specific class of reasons for scientific behaviour that are not dependent upon general axiology.

Without going into too much detail, we can say that Feyerabend's account of the physical grounding of complementarity works out Bohr's assumption of the *indeterminateness of state descriptions*, of which he takes complementarity to be an abstract generalisation. He tracks in detail the introduction of the assumption as a 'physical hypothesis' (he underlines time and again its objective character)[10] to make the gradual interaction between two physical systems consistent with the quantum postulate: 'during the interaction between two systems (A) and (B), the dynamical states of both (A) and (B) cease to be well defined so that it becomes *meaningless* (rather than false) to ascribe a definite energy to either of them' (p. 196). His point, which he makes time and again, is to bring out the objective character of this 'simple and ingenious physical hypothesis', which is only based on the quantum postulate and duality (together with the individual conservation of energy and momentum, cf. Feyerabend 1962b, p. 204): indeterminateness is introduced by Bohr not on the basis of a verificationist theory of meaning (though he admits it has been used in this connection by many other physicists and philosophers), but on the basis of 'well-known classical examples of terms which are meaningfully applied only if certain physical conditions are first satisfied and which become inapplicable and therefore meaningless as soon as the conditions cease to hold.'[11]

Second, he calls out the misconception that Bohr's hypothesis makes reference to knowledge or observability; as a physical hypothesis, it 'excludes' the *existence* of 'these intermediate states themselves' (p. 197). Having dispelled misreadings of the indeterminateness assumption *as proposed by Bohr*, he proceeds to explain how the hypothesis stood up successfully against two alternatives (Planck and Schrödinger: ψ-waves as complete and well-defined states; statistical ensemble interpretation of the ψ-function) to explain the physical and conceptual problems on the table (pp. 203–207). He shows how Bohr himself tried to come up with alternatives, only to be thrown back to indeterminateness as the only viable solution.

As a preliminary conclusion, he points out that it is 'impossible to derive Bohr's hypothesis ... from the formalism of the wave mechanics plus the Born interpretation' (p. 207), and, since the 'qualitative considerations'

---

[10] Cf. Feyerabend (1962c, p. 194, p. 220, emphasis added): 'Most critics interpret the two main principles of the Copenhagen Interpretation, namely, the principle of the indeterminateness of state descriptions and the principle of the relational character of quantum mechanical states not as physical assumptions which describe *objective features of physical systems*; they interpret them as the direct result of a positivistic epistemology and reject them together with the latter. [...] We first presented a physical hypothesis which was introduced by Bohr to explain certain features of microscopic systems (for example, their wave properties). It was pointed out that this physical hypothesis is of a *purely objective character* and that it is also needed, in addition to Born's rules, for a satisfactory interpretation of the formalism of wave mechanics.'

[11] In this connection he uses the example of the term 'scratchability' (Mohs' scale of mineral hardness) 'which is applicable to rigid bodies only and which loses its significance as soon as the bodies start melting' (Feyerabend 1962c, p. 197). (This example is repeatedly used to suggest a non-philosophical reading, see 1958b, p. 51; 1960c, p. 323; 1961, p. 373; 1964, p. 294; 1969b, pp. 94–95).

behind the hypothesis 'are needed in addition to Born's interpretations if a full understanding of the theory [i.e. the formalism of wave mechanics] is to be achieved', Born's hypothesis of the indeterminateness of state descriptions is an irreducible, i.e. independent *and* necessary part of quantum mechanics (p. 208).

Feyerabend (pp. 208–220) then proceeds to explain how Bohr's second hypothesis, the assumption of the *relational character of quantum-mechanical states*, was proposed as a response to Einstein, Podolsky and Rosen (1935) and how it is intimately connected to the first hypothesis insofar as it grew out of the same qualitative considerations that brought about indeterminateness. (In this sense it is not an ad hoc move to accommodate the 'very surprising case discussed by EPR' (Feyerabend 1962b p. 218)). Instead of assuming, as EPR had done, that 'what we determine when all interference has been eliminated is a property of the system investigated', Bohr maintains that 'all state descriptions of quantum mechanical systems are *relations* between the systems and measuring devices in action and are therefore dependent upon the existence of other systems suitable for carrying out the measurement' (p. 217). This is the second hypothesis. Feyerabend goes on to show 'how this second basic postulate of Bohr's point of view makes indefiniteness of state descriptions compatible with EPR. For while a property cannot be changed except by *interference* with the system that possessed that property, a relation can be changed without such interference' (p. 217).

Finally, Feyerabend introduces Bohr's principle of complementarity. Where the indeterminateness hypothesis referred to 'description in terms of classical concepts and asserted that description in terms of these concepts must be made 'more liberal' if agreement with experiment is to be obtained', this principle 'expresses in more general terms this restriction, forced upon by experiment, in the handling of the classical concepts' (p. 222). To show in which way our interpretation of Feyerabend's view that the complementarity principle 'had earned its place in microphysics' holds, we have to carefully disentangle Feyerabend's discussion of complementarity. The complementarity principle is not identical with the indefiniteness hypothesis, it is a philosophical extension. Empirically, it assumes (beside the conservation laws) duality and the quantum of action, but it also introduces 'some further premises which are neither empirical, nor mathematical, and which may therefore be properly called "metaphysical"' (p. 222). Because Feyerabend uses the rest of the paper to severely criticise these further assumptions from a methodological point of view, it may seem that he does *reject* complementarity after all. But this is not correct. First, we have to distinguish Feyerabend's recognition that complementarity (i.e. including these metaphysical assumptions) has 'earned its place in microphysics' in that its application in microphysics was successful in advancing its development: the existence of quantum mechanics vindicates the abstract principle of complementarity. Feyerabend is very clear on this point when he discusses how the more 'liberal attitude towards' classical concepts had been guided by the correspondence rule to obtain 'rational [or natural] generalization of the classical mode of description':

> [I]t is very important to realise that a 'rational generalization' ... does not admit of a realistic interpretation of any of its terms. The classical terms cannot be interpreted in a realistic manner as their application is restricted to a description of experimental results. The remaining terms cannot be interpreted realistically either as they have been introduced for the explicit purpose of enabling the physicist to handle the classical terms properly. The instrumentalism of the quantum theory is therefore *not a philosophical manoeuvre that has been willfully superimposed upon a theory which would have looked much better when interpreted in a realistic fashion. It is a demand for theory construction which was imposed from the beginning and in accordance with which, part of the quantum theory was actually obtained.* (p. 265 fn. 62)

But complementarity, as a general principle of Bohr's Copenhagen interpretation, claims validity beyond quantum mechanics. While Feyerabend even agrees that its success may warrant complementarity as a useful heuristic principle for future development, he understands Bohr to make a much stronger claim: any future microphysical theory that will not obey complementarity

> will either be internally inconsistent, or inconsistent with some very important experimental results. [Many followers of the 'orthodox' point of view] therefore not only suggest an interpretation of the known results in terms of indefinite state descriptions. They also suggest that this interpretation *be retained forever* and that it be the foundation of any future theory at the microlevel. It is at this point that we shall have to part company. I am prepared to defend the Copenhagen Interpretation as a physical hypothesis and I am also prepared to admit that it is superior to a host of alternative interpretations ... But ... any argument that wants to establish this interpretation more firmly is doomed to failure. (p. 201)

Thus, Feyerabend rejects the complementarity principle insofar as it implies that its success in the construction of quantum mechanics warrants its extension to any future microphysical theory, that is, its imposition as a *necessary restriction* on the future development of physics. Additionally, he rejects complementarity on general methodological grounds, greatly expanding on his arguments concerning complementarity as a new model of explanation already discussed above (cf. 1958c, p. 90) and to be further discussed below.

More generally, behind this re-evaluation of Bohr's arguments lies Feyerabend's consistent aim to understand Bohr's thinking as original contributions to quantum theory, not at all assimilable to other members of the Copenhagen school. In this respect, we must call into question Don Howard's claim that Feyerabend was among 'the most important enablers of the myth' (Howard 2004, p. 677) of a unitary Copenhagen interpretation allegedly reproducing Bohr's view; if Feyerabend was part of the group who most 'contributed to the promotion of this invention for polemical or rhetorical purposes' (p. 670), this claim should be limited to his pre-1958 papers. After this, he was an active myth-buster.

7.4 Physical Arguments and Ontological Problems

Let us dwell a little longer on the new and remarkable outcome of Feyerabend's investigation: Bohr's interpretation, in particular the principle of complementarity, is justified by physical arguments grounded in Bohr's research activity.[12] This, however, seems at odds with the contention, stemming from Feyerabend's philosophical prescriptivism, that the interpretation of quantum theory is a philosophical problem to be decided on purely methodological grounds. Has the interpretation of quantum theory suddenly become a physical question? To understand how Feyerabend understood this state of affairs, it is instructive to see how he conceptualised the interplay between philosophical and physical problems in the domain of quantum mechanics.

In a letter to Herbert Feigl from 28 June 1957, a few months after the Colston Symposium, Feyerabend sketched a framework for the discussion of quantum mechanics for an upcoming conference to be held at the Minnesota Centre for the Philosophy of Science. First, he drew the distinction between the 'analysis of quantum mechanics in its present form and interpretation' and 'suggestions as to the possible form of a future theory of microscopic phenomena' (Feyerabend to Feigl, 28 June 1957, HF 02–133–02/1). This was by no means an obvious distinction at the time. Let us remember: the completeness of quantum theory was assumed; and, as Leon Rosenfeld did, the very expression 'interpretation' was questioned because the term

---

[12] The interpretation may have had roots in Bohr's philosophical ideas, and Feyerabend is not disputing this. Feyerabend's point is that the interpretation earns its place in physics not because of its philosophical background, but because of Bohr's physical arguments.

suggested that other interpretations were possible. Second, he distinguished between 'syntax' and 'semantics' of elementary quantum theory, that is, the chosen mathematical formalism and the rules, which are 'necessary and sufficient for transforming the formalism into a full-fledged physical theory'. Notably, questions about the proper interpretation of quantum mechanics are not semantical questions, but take place on a third level, 'ontology':

> [W]hen discussing the question which is the proper interpretation of quantum mechanics, a wave-interpretation, a particle interpretation or e.g. the Copenhagen-interpretation, physicists and philosophers are not concerned with semantical problems, i.e. they are not concerned with the problem how an uninterpreted formalism ought to be connected 'with reality'. The question 'particles or waves?' rather presupposes that the symbols of quantum mechanics have already been given a certain meaning, i.e. it presupposes that all syntactical and semantical problems have been settled in a satisfactory way. What is to be interpreted is not a formalism, but a physical theory. This is the reason why it seems to be advisable to distinguish between two different kinds of interpretation of a physical theory, between its *semantical interpretation* and its *ontological interpretation*. The Born-interpretation is a semantical interpretation of the formalism of quantum mechanics. The Copenhagen-interpretation (or the wave-interpretation or the particle-interpretation) is an ontological interpretation of quantum theory. Problems connected with ontological interpretations I shall call ontological problems. This distinction between syntactical problems, semantic problems, ontological problems, seems to be very useful, especially in the case of quantum mechanics. (Feyerabend to Feigl, 28 June 1957, HF 02–133–02/1)

Among the ontological interpretations of quantum theory, Feyerabend lists Einstein's – 'as defended by Popper'; Bohm's first (1952) interpretation; similarity between quantum mechanics and the theory of diffusion; and Schrödinger's interpretation (Feyerabend to Feigl, p. 3). As alternative theories to quantum mechanics, with their own possible sets of ontological interpretations, Feyerabend mentions 'Bohm's new papers'.[13]

This very specific organisation of the discussion has a number of consequences relative to how the levels are related to each other. With regard to the ontological level, Feyerabend is very clear that there can be a relation of implication between this level and the syntactic plus semantic level:

---

[13] Presumably Bohm (1953), Bohm and Vigier (1954), and possibly Bohm and Aharonov (1957); Feyerabend might also have had Bohm's Colston Symposium paper (Bohm 1957a) in mind; see also Bohm (1957b), which, though not a paper, Feyerabend was already acquainted with in April 1957 at the latest (cf. Feyerabend to Popper, 1 April 1957, KP in Feyerabend 2020, p. 259).

Traditional philosophers have tried to solve ontological (or metaphysical) problems such as e.g. the problem of causality (or the narrower problem of determinism) by speculation on the basis of (sometimes very scarce) experience. The existence of very general scientific theories enables the philosopher to change the methods of ontological research. For it may turn out that a theorem of one of those theories either contradicts, or implies a statement of metaphysics. Such a theorem may be called 'ontologically relevant'. *And a hypothesis as to the ontologically relevant theorems of a given theory may be called an ontological interpretation of that theory.* Ontological interpretations in this sense can be *tested* by comparing their consequences with theorems of the theory so interpreted. It is not always easy to carry out such a test. This is the reason why there is still so much *argument* about the (ontological) interpretation of quantum-mechanics. On the other hand [ontological interpretations] may be introduced with the help of certain arguments which do not at all refer to theorems of the theory so interpreted and which strongly resemble the ontological arguments of traditional metaphysics. Most of Bohr's arguments are of this kind, although his results are shown to be correct by many theorems of the theory itself. (Feyerabend to Feigl, 28 June 1957, HF 02–133–02/1, emphasis in the original)

Note that this is the state of Feyerabend's assessment in 1957, that is, this framework is in place *before* Feyerabend's re-evaluation of Bohr. This tells us two things: first, the contention that general physical theories are relevant to ontological problems that were once in the domain of 'pure metaphysics' (e.g. the issue of determinism) precedes the re-evaluation of Bohr. (Indeed, this contention is one of the most pristine expressions of Feyerabend's understanding of the rapprochement of science and philosophy and it was already clearly expressed in Feyerabend (1954b/2015)).

Second, he still thought that the 'Copenhagen interpretation', including Bohr's complementarity principle, was not an ontological interpretation derived from the underlying physical theory, but – following Popper's assessment – was posited on the basis of a dubious philosophical presupposition. Feyerabend recognised that it fit the underlying physical theory, but it was almost as if it matched it 'by chance'. This coincides with the outline of Feyerabend (1958c) that we gave above, in particular point (a).

Feyerabend's re-evaluation of Bohr's ontological interpretation as being grounded in physical argument (1962b) does not overthrow this framework in principle. Indeed, such a move is envisaged in the framework and corresponds to the possibility that 'a theorem of one of those theories either contradicts, or implies a statement of metaphysics'. Feyerabend's claim that Bohr's 'point of view can stand upon its own feet and does not need any support from philosophy' is equipollent to the claim that it is an 'ontologically relevant' consequence of the physical theory, not a philosophical argument 'resembling ontological arguments of traditional metaphysics', as previously thought. And yet, behind this coherent interplay of philosophy and science lurks a possibility that Feyerabend had not readily envisaged. When Feyerabend thought of ontological problems, he thought of issues like determinism. But the upshot of his reevaluation of Bohr is that another kind of issue turns out to be an 'ontologically relevant' consequence of physical theory: the issue of realism itself. This result cannot be overstated: under the assumption that realism and instrumentalism are mutually excluding positions, we have a situation in which

1. according to general axiology, there are compelling reasons to interpret scientific theories realistically;
2. the instrumentalist interpretation of a specific theory, namely quantum mechanics, is compelling because of physical arguments grounded in the development of the theory.

In other words, we are now confronted with an explicit case of Feyerabend's dilemma.

How did Feyerabend deal with the dilemma? Quite ingeniously, he used his theoretical pluralism to give an answer: while a given physical theory (its syntax and semantics) may indeed give stringent indications as to the right solution to an ontological problem, including the realism issue, a methodological demand can always be put forward to develop genuinely alternative theories that may imply different solutions to ontological problems. If quantum mechanics forces an instrumentalist interpretation, the importance of genuine alternatives to quantum mechanics that allow for a realistic interpretation becomes a central problem for the future of microphysics.

7.5   The Limits of Quantum Theory and Hidden Variable Alternatives

At first, Feyerabend made the contextual re-evaluation of complementarity fit with general methodology, in particular with his methodological argument for realism presented in the introductory section. If not only elementary quantum mechanics but also Bohr's interpretation had earned their right to stay, it was not the interpretation that was in need of being changed:

> If I am correct in this, then all those philosophers who try to solve the quantum riddle by trying to provide an alternative interpretation of the

current theory which leaves all laws of this theory unchanged are wasting their time. Those who are not satisfied with the Copenhagen point of view must realise that only a new theory will be capable of satisfying their demands. (Feyerabend 1962b, p. 260 fn. 29)

Progress could only come from an alternative, realistically formulable theory, whose purpose was to compete with quantum mechanics in the microphysical domain and, while being in accordance with quantum mechanics to an approximation, would *contradict* quantum theory.

This point had been made already long before his reappraisal of Bohr while discussing how a future microphysical theory should look like (still assuming von Neumann's no-go theorem to be unlimitedly valid).[14] The philosophical outline about how a new theory in the microphysical domain should look like was a direct application of the anti-incrementalist notion of historical progress of theory succession qua theory replacement. This is one of the most durable notions throughout Feyerabend's philosophical papers. Not so well known is that its genesis and argumentative use starts out in his papers on quantum mechanics, in which he consistently referred to the historical example of the inter-theoretic relation between Kepler's and Newton's laws and referenced (often, but not always) a little-known paper by Karl Popper (1949).[15] The first use of the Kepler–Newton transition happens while discussing the question whether Bohm's first attempt (1952) at a hidden variables interpretation could bring back determinism in the realm of quantum theory.[16] Feyerabend's conclusion is that it cannot in its current form, but the reason lies in the fact that 'Bohm takes up the task to construct an interpretation that does *not* contradict quantum theory':

> Physicists and philosophers who defend the idea that a causal interpretation of the formulas of quantum mechanics is possible are always very concerned that this interpretation does not contradict quantum theory. That is why von Neumann's proof seemed, for them, to represent an obstacle that could not be overcome. As a consequence, they overlook the fact that

[14] And indeed von Neumann (1932/1955, p. 325) made the same point when he commented on his proof: 'we need not go any further into the mechanism of the "hidden parameters", since we know that the established results of quantum mechanics can never be re-derived with their help. . . . The present system of quantum mechanics *would have to be objectively false*, in order that another description or the elementary processes than the statistical one may be possible' (emphasis added) – Feyerabend 'simply' added the methodological justification to pursue this goal.

[15] This chapter is of some historical significance for Feyerabend scholarship. It is the paper Popper gave at the *Internationalen Hochschulwochen* at Alpbach in 1948, when Feyerabend first met Popper; see Kuby (2010) for details. The paper appeared in English translation in Popper (1963). Popper repeated the point in his (1983, p. 140), which is now the *locus classicus*.

[16] Bohm regarded his 1952 proposal as a proof-of-concept to show the limit of von Neumann's no-go theorem and thus the *possibility* of a hidden variables approach, not as an alternative physical theory.

comprehensive theories, which unify a series of less comprehensive theories, almost invariably contradict them: Kepler's laws contradict Newton's theory, as they can be derived from it only approximately. As a consequence, as long as the contradiction between quantum theory and its allegedly causal interpretation falls under the threshold of measurement, its existence cannot be used as an argument against the interpretation. (Feyerabend 1954a/2015, pp. 39–40; cf. Feyerabend 1954a, 104)

The historical point is repeated time and again from Feyerabend (1954b, pp. 470–471)[17] to Feyerabend (1965, p. 236 fn. 44); the argumentative move can be found in Feyerabend (1958a).[18] And this point was not only made by Feyerabend on behalf of 'quantum dissidents', but was made by the dissidents themselves. Already at the Colston Symposium Bohm is recorded as saying:

> I agree with Professor Rosenfeld that our theory cannot be entirely equivalent to quantum mechanics, but I also believe that every new theory must contradict the old theory in some respects. Quantum mechanics contradicts classical mechanics in very important respects ... and nevertheless approaches classical mechanics as an approximation ... I believe that eventually we will come to a point where we contradict quantum mechanics and get consequences which simply are not consistent with the quantum of action. (Körner 1957, p. 46)[19]

As we can see, Feyerabend's appreciation for quantum theory and Bohr's interpretation, on the one side, and his interest in alternative microphysical theories, on the other side, was not in contradiction; it was part of one and the same research problem: the question how a real alternative to quantum mechanics looks could be answered by studying the limits of quantum theory. (Feyerabend 1965a, p. 251 fn. 125)

7.6 The Problem of Competing Methodological Rules

Is Feyerabend's way of disengaging from the dilemma appealing? In part, it is: as the notion of progress through theoretical pluralism was built

---

[17] 'The movement of the elements is very well described by Kepler's laws. However, these laws contradict Newton's theory (for they are valid only for an infinitely heavy Sun and for the planets with negligible masses)' (Feyerabend 1954b/2015, p. 17; cf. Feyerabend 1954b, pp. 470–471).

[18] '[E]ven if (a) and (b) were theorems of [quantum mechanics] von Neumann's proof could not show, as has sometimes been assumed, that determinism has been eliminated once and forever. For new theories of atomic phenomena will have to be more general; they will contain the present theory as an approximation; which means that, strictly speaking, they will contradict the present theory. Hence, they need no longer allow for the derivation of von Neumann's theorem' (Feyerabend 1958a, p. 345).

[19] The point that not every future microphysical theory will need to accommodate Planck's constant 'in an essential way' is repeated in Feyerabend (1962c, p. 227).

independently of the dilemma, developing it further to dissolve the dilemma doesn't seem an ad hoc move to save general methodology as a justification for axiological arguments for realism. Instead, it can be used as a further reason in an argument about the progress of science. This argument was the development of theoretical pluralism as a methodological proposal, which had been in the making for some time.

And yet – and here I introduce the problematic kernel – divorcing the future progress of physics and the development of quantum mechanics (which after all was the future of physics at some point in time!) obscures the incompatibility of two opposite methodological rules applying to one and the same situation. Assume we take complementarity to be a methodological principle about how to handle statements involving classical concepts; then this principle, which tells scientists to restrict the validity of these statements, directly contradicts the methodological rule following from the principle of testability, according to which scientists ought to force the universal validity of these statements. And, in the specific instance of the development of quantum mechanics, Feyerabend is ready to admit that complementarity trumps a realistic interpretation, that is an instrumentalist interpretation is the 'right scientific move', the justified behavioural guideline as a mean to realise the principle of testability. Following our analysis of two levels of complementarity, Feyerabend circumvents the problem described because he avoids a methodological reading of complementarity grounded in physical argument on the one side, and rejects complementarity when viewed as a generalisation justified on philosophical grounds on the other side. The strong emphasis on the physical grounding of complementarity has a double argumentative function. Since following his philosophical prescriptivism, physical reasons cannot justify general methodological rules and only axiological decisions can, as long as complementarity is treated as physically justified, it cannot have the status of a methodological rule – this avoids having to describe the situation of quantum mechanics in a way in which two general methodological rules, both justified on quite different grounds, are in conflict; and, where it is extended by further philosophical reasons to become a methodological rule, the philosophical reasons adduced can be thoroughly criticised on axiological grounds and are shown to be 'neither correct nor reasonable' (Feyerabend 1962b, p. 195).

This leads to a very interesting if unintended result: Feyerabend's construal and appreciation of complementarity as a 'mere' heuristic move grounded in a specific research situation is actually the first instance of what he would later call a 'rule of thumb': in contrast to methodological rules, there is no general justification for its application, it is only contextually valid in the scientific situation in which it shows its worth, and its future success cannot be inferred from its past success. Feyerabend's refusal to elevate complementarity to a general methodological rule (or to interpret indeterminateness as an application of this methodology) provides the template for his later negation of the existence of general methodological rules.

There is, furthermore, an even more obvious candidate for a methodological rule, the correspondence rule, which, as Feyerabend himself reports, is a '*demand for theory construction* which was imposed from the beginning and in accordance with which, part of the quantum theory was actually obtained' (p. 265 fn. 62, emphasis added). In this case, Feyerabend disengages the threat by limiting its reach, for it did not bring about the 'other part' of quantum theory: wave mechanics. Wave mechanics as the completion of quantum mechanics was, instead, constructed following a realistic demand 'that was completely opposed to the philosophical point of view of Niels Bohr and his disciples' (p. 265 fn. 62) including the correspondence rule. That wave mechanics turned out to be 'just that complete rational generalization of the classical theory that Bohr, Heisenberg and their collaborators had been looking for' (p. 265. fn. 62) is thus a lucky coincidence, not a result attributable to the correspondence rule. Similarly, to the complementarity case, this handling of correspondence is a preview of a later concept, the notion of the limited validity of methodological rules.

Both cases show *in nuce* the difficulties that eventually would motivate Feyerabend to drop the universal justifiability of methodological rules. However, it does not need an incompatible methodological rule to provide a counter-instance to a given methodology. Feyerabend already admits that complementarity earned its place because a realistic interpretation didn't work out notwithstanding many attempts in this direction (also by Bohr, contrary to the latter's own philosophical inclinations):

> [T]he [preceding] arguments . . . should have shown that there exist weighty *physical* reasons why at the present moment a realistic interpretation of the wave mechanics does not seem to be feasible . . . A philosophical crusade for realism alone will not be able to eliminate these arguments. At best, it can ignore them. What is needed is a new theory. Nothing less will do. (p. 260 fn. 49)

This negative result of achieving a realistic interpretation directly impinges on the *realisability* of Feyerabend's methodological proposal. But, as is well known, a counter-instance does not make a falsification and Feyerabend is adamant that the 'failure' of his methodological rule in a specific instance does not prove that it cannot be successful in the future (Feyerabend's papers abound with syntactical double negation constructions in this regard), which amounts to the assertion that an alternative to quantum theory allowing for a realistic interpretation has not been shown to be impossible. Feyerabend continues the preceding quotation:

> I have to admit, however ... that philosophical arguments for realism, though not sufficient, are therefore not unnecessary. It has been shown that given the laws of wave mechanics, it is impossible to construct a realistic interpretation of this very same theory. That is, it has been shown that the usual philosophical arguments in favour of a realistic interpretations of theoretical terms do not work in the case of quantum mechanics. [T]here still remains the fact that theories which *do* admit of a realistic interpretation are definitely preferable to theories which do not. It was this belief which has inspired Einstein, Schrödinger, Bohm, Vigier and others to look for a modification of the present theory that makes realism again possible. The main aim of the present article is to show that there are no valid reasons to assume that this valiant attempt is bound to be unsuccessful. (p. 260 fn. 49)

This sounds like all is well on the philosophical battlefield, but in fact this is a retreat. Feyerabend moves the goalpost from a methodological assurance that a realistic theory is not only desirable but realisable to the claim that such a theory has not been shown to be impossible or that the attempt to find one will be unsuccessful. We want to draw attention to this shift because there is a lesson to be learned from his re-evaluation of Bohr: the realisability of a realistic interpretation is not a given.

Going back to Feyerabend's methodological arguments for realism and proliferation, we discover a further (necessary but unstated) premise, that scientific theories are in principle amenable to a realistic interpretation. This premise turns out to be false. The premise is quite innocent under the assumption that a realistic interpretation depends only on a decision about how to handle scientific statements, a decision independent from physical results, and the specifics of the theory we want give a realistic interpretation of. But now, it turns out that the specifics of actual research can pose constraints on this handling. *This is the moment in which justified actual scientific practice comes in contact – one may say: comes in the way of – Feyerabend's conception of methodology as conceived in his philosophical prescriptivism.*

Feyerabend found himself in a tough spot. He welcomed cases in which philosophical notions come in contact with experience; at the same time, he needed the philosophical notion of realism to be a consequence of volitional decisions. His response was ambivalent: he recognised the result, but he did not accept the consequences, making several attempts not to give up the methodological argument for a realistic interpretation of alternative theories by bringing the principle of testability to its argumentative limits. The best example is his paper 'Realism and Instrumentalism: Comments of the Logic of Factual Support' (1964), which, notably, is devoted to flatly arguing 'that realism is preferable to instrumentalism' (Feyerabend 1964, p. 280). He further strengthened the argument for proliferation by claiming that alternative theories are not only more likely to maximise the testability of established theories, but also that there exist situations in which realistically interpreted alternative theories are necessary in principle to test the established theory. But this argument chokes in light of his re-evaluation of Bohr. The methodological arguments for realism work only as long as we disregard the (admittedly surprising) discovery that the issue of ontological interpretation can be an ontologically relevant consequence of physical theory. For it is now possible that no future theory will admit a realistic interpretation on scientific grounds. His methodological argument for a realistic interpretation of scientific theories has become unsuccessful.[20]

The argument's failure is not a black box. We can pinpoint the exact source of the problem: it lies in the fact that the principle of testability cannot warrant an inference to realism anymore, which is the very core of all Feyerabend's methodological arguments for realism. We can also speculate as to why Feyerabend did not immediately recognise this problem. The contextualisation of an argument for realism in a broader argument for theoretical pluralism put the development of alternative theories at the centre of attention: this was an independent mean to realising the testability principle; the realistic interpretation of these alternatives has become an additional step towards testability. And Feyerabend was right to push his argument insofar as the argument for theory proliferation (as distinct from their realistic interpretation) still works; it remains unaffected by the discovery that the issue of interpretation can be amongst the ontological consequences of a scientific theory. But Feyerabend wanted more. As late as the date of the paper under scrutiny, Feyerabend thought that also the demand of a realistic interpretation of those empirically (still) unconfirmed

---

[20] Barring, that is, the discovery of a principle applicable to theory construction that can guarantee a realistic interpretation on physical grounds in addition to all other requirements that a successor theory has to fulfill; or a realisability condition that can guarantee that issues of interpretation are excluded from the relevant ontologically consequences of the theory so constructed. None of these options have been explored by Feyerabend, as far as I am aware.

alternatives was a plausible demand which immediately follows from the principle of testability (1964, p. 308). But this further inference is now unwarranted. As he lays down his argument, Feyerabend even distinguishes the two points:

> [1:] the development of such further theories is demanded by the principle of testability, according to which it is the task of the scientist relentlessly to test whatever theory he possesses [2:] and it is also demanded that these further theories be developed in their strongest possible form, i.e. as descriptions of reality rather than as mere instruments of successful prediction. (p. 306)

In this passage, we can pinpoint where Feyerabend's principle of testability as a cogent argument for realism chokes in light of his re-evaluation of Bohr: (1) still works, but (2) does not. In other words, Feyerabend did distinguish the two points, but did not distinguish their different warrants.

7.7 Conclusion

The strong methodological argument for realism fails, and I claim that Feyerabend came to this realisation, too.[21] The conceptual problem behind the argumentative failure lies in the equation of 'in their strongest form' and 'descriptions of reality' to mean 'realism'. This very equality has been shown to be wrong by his re-evaluation of Bohr. The discovery that the issue of realism itself can be an 'ontologically relevant' consequence of a physical theory is not only potentially disruptive vis-à-vis axiological arguments for realism, it leads by itself to an almost paradoxical situation in the case of quantum mechanics:

1. Realism exhorts scientists to take the ontological consequences of their physical theories at face value (to develop them 'in their strongest possible form').
2. Taking the ontological consequences of quantum mechanics at face value (to develop them 'in their strongest possible form') results in an instrumentalist interpretation of the theory.

If the expression 'to take the ontological consequences of a physical theory seriously' was used by Feyerabend synonymously with a realistic

---

[21] His review of Ernest Nagel's *Structure of Science* (Feyerabend 1966) is the last published appearance of the argument that 'strong reasons' against a realistic interpretation of the quantum theory 'can be removed only by arguments showing that it is desirable to introduce theories which contradict already existing laws' and he shows 'that such arguments can be provided' (p. 248). All later references to a strong methodological argument for theoretical pluralism only concern the proliferation principle proper; the realistic interpretation of alternatives is now omitted.

interpretation of the theory, in the specific case of quantum mechanics, it leads to the opposite interpretation, in that it forces us to accept the limited validity of central concepts of the theory, as Bohr had argued. This is not an instance of Feyerabend's dilemma (which concerned competing sources of justification of how to interpret scientific theories), but shows a problem with Feyerabend's realism, that is with the concept of 'ontological inter-pretation' of a scientific theory itself, which escaped him at first, perhaps because of his thinking in Popperian terms of 'positivism' versus 'realism'. Feyerabend's tendency to describe the interpretative situation of quantum mechanics in a (grammatically and evaluative) negative way, that is as the 'impossibility' of a realistic interpretation, forbade him from appreciating Bohr's interpretative outcome as a fully fledged ontological interpretation, that is, instrumentalism as possible 'description of reality'. The discovery, in the end, amounts to a refutation of Feyerabend's philosophical concept of realism in its general application to science, that is it shows the inadequ-ateness of hidden philosophical premises in Feyerabend's realistic conception.

Feyerabend came not only to recognise this point, indeed he embraced it. In his introduction to the publication of his *Collected Papers*, Volume I, he commented on two reissued papers (including the paper discussed at length in this section) by admitting that, because of the specific arguments found in his re-evaluation of Bohr, these turn out to be 'somewhat mis-leading' (Feyerabend 1981b, p. 15):

> Producing philosophical arguments for a point of view whose applicability has to be decided by concrete scientific research, they suggest that scientific realism is the only reasonable position to take, come what may, and inject a dogmatic element into scientific discussion ... Of course, philosophical arguments should not be avoided; but they have to pass the test of scientific practice. *They are welcome if they help the practice*; they must be withdrawn if they hinder it, or deflect it in undesirable directions. (pp. 15–16)

> The issue between realism and instrumentalism gives rise to similar observa-tions. Do electrons exist or are they merely fictitious ideas for the ordering of observations (sense data, classical events)? It would seem that the question has to be decided by research. ... Modern professional realists do not see matters in this way. For them, the interpretation of theories can be decided on purely methodological grounds and independently of scientific research. Small wonder that their notion of reality and that of the scientists have hardly anything in common. (Feyerabend 1978b, p. 39)

A consequence of this new view applied to realism is first presented in his paper 'On a Recent Critique of Complementarity' (1968a; 1969b).

Prompted by widely received critiques of the Copenhagen interpretation by Mario Bunge and Karl Popper in Bunge (1967), Feyerabend reissued once more his arguments about the physical grounding of Bohr's point of view. But this time, he did not attempt to limit its ontological consequences on methodological grounds. A methodological argument for realism was nowhere to be found. Instead, he exposed 'the myth of Bohr's dogmatism' (Feyerabend 1969b, p. 85 fn. 61), pointed out Bunge's ignorance 'of Bohr and the actual development of ideas within the 'Copenhagen Circle'' (p. 92 fn. 81) and explained how Bohr's interpretation had arisen from a process of 'refutations and discoveries', not of 'philosophical dogmatism' (p. 92). Feyerabend's conclusion now was 'back to Bohr!' (p. 103).[22]

[22] Acknowledgements: I am thankful to Carolin Antos, Neil Barton, Matteo Collodel, Michael Heidelberger, Paul Hoyningen-Huene, Deborah Kant, Jamie Shaw, as well as my Ph.D. advisors Elisabeth Nemeth and Wolgang Reiter for critical feedback on earlier presentations and drafts of this chapter. I received valuable comments in Martin Kusch's colloquium at the University of Vienna, in which I presented a related paper. Only recently it came to my attention that Marij van Strien has been working on a paper (van Strien 2019) that connects Feyerabend's technical work on quantum mechanics and his general philosophy of science in a similar way. I am thankful to her for correspondence about the similarities and differences of our views.

The archival documents cited in this chapter are published (many for the first time) in Feyerabend (2020). I thank the editors Matteo Collodel and Eric Oberheim for making this material available to me before publication.

This work has been funded in part by the Austrian Science Fund (FWF W1228-G18) and in part by the Volkswagen Foundation (Project "Forcing: Conceptual Change in the Foundations of Mathematics").

CHAPTER 8

On Feyerabend, General Relativity, and "Unreasonable" Universes

J. B. Manchak

8.1 Introduction

In what follows, I will investigate the principle anything goes within the context of general relativity. It is of some interest that even after one restricts attention in this way, one can still carry out a sort of Dadaist "joyful experiment" (Feyerabend [1975] 2010, xiv) demonstrating the chimerical nature of various distinctions between 'reasonable' and 'unreasonable' models of the universe.[1] Here, I intend to sketch the contours of such an undertaking without presupposing any prior familiarity with general relativity on the part of the reader.

In the first portion of the paper, I will walk through some of the basic structure of general relativity. I then consider a remark made by Feyerabend in the introduction to *Against Method*. There, we are told that a pluralistic approach to epistemology seems to be appropriate given that "the world which we want to explore is a largely unknown entity" (Feyerabend [1975] 2010, 4). The meaning of the claim is not clarified and no argument is ever given for it. But, if true, it does seem to lead naturally to the position that we ought "keep our options open" with respect to competing models of the universe (Feyerabend [1975] 2010, 4). In the second portion of the paper, I will articulate a sense in which the claim is true with respect to the models of the universe compatible with general relativity; any idealized person represented in virtually any such model will not have the epistemic resources – even with a robust type of inductive reasoning – to know which model she inhabits. For her, the

---

[1] From 1970 until 1988, both the paper and book versions of *Against Method* were given a subtitle – Outline of an Anarchistic Theory of Knowledge – and an accompanying footnote in which Feyerabend distanced himself from the 'seriousness' of anarchism and asked that he be remembered as a Dadaist instead. For more on Feyerabend and Dada, see the introductory note by Ian Hacking in the fourth edition of *Against Method* (Feyerabend [1975] 2010, xiii–xvi).

universe is (and will always remain) a largely unknown entity. I will then consider the esteemed property of inextendibility which requires that a model universe be 'as large as it can be' in a natural sense. This property is almost universally thought to be satisfied by all 'reasonable' models of the universe. But I will emphasize that any idealized person in virtually any model cannot know that her universe is inextendible. The upshot is this: within this context, there is (and will always remain) the epistemic possibility that our universe is best represented by an 'unreasonable' model.

Given the state of affairs, proceeding counter-inductively seems to be especially appropriate.[2] In particular, the practice of making "the weaker case the stronger" (Feyerabend [1975] 2010, 14) can be used to blur some of the usual lines between 'reasonable' and 'unreasonable' models of the universe. In the third portion of the paper, I will return to the property of inextendibility and work to cast doubt on the idea that a 'reasonable' model universe must be 'as large as it can be.' I will do this in two steps. First, I will argue that the usual distinctions between 'reasonable' and 'unreasonable' models can be upheld only if the definition of inextendibility is radically modified; the now standard formulation allows for extendible models to nonetheless be 'as large as they can be' in the sense that they cannot be 'reasonably' extended. Second, I will call into question the metaphysical underpinnings of the position that our universe is 'as large as it can be' in light of the need to modify the definition of inextendibility. A celebrated foundational result states that every extendible model of the universe has at least one corresponding inextendible extension. But I will show that under some 'reasonable' revisions to the definition of inextendibility, the analogous results do not hold; it is not always possible for a 'reasonable' model of the universe to be 'as large as it can be.'

Along the way, we seem to be led to consider an "ocean of mutually incompatible alternatives" to the standard formulation of general relativity (Feyer-abend [1975] 2010, 14). I will close with an articulation of such alternatives – each a variant of general relativity incompatible with

---

[2] One may proceed counter-inductively by using "hypotheses that contradict well-confrmed theories and/or well-established facts" (Feyerabend [1975] 2010, 13). In what follows, I will only consider hypotheses that contradict well-confrmed theories rather than well-established facts. In this sense, I will be also be following Feyerabend's earlier principle of proliferation: "Invent, and elaborate theories which are inconsistent with the accepted point of view, even if the latter should happen to be highly confirmed and generally accepted" (Feyerabend 1965, 101). For more on the principle of proliferation, see Shaw (2017).

the standard formulation. I hope to show that by contrasting the situation in one variant theory with another, we can come to understand foundational questions within 'general relativity' in a more nuanced way. I will sketch some of the work ahead if we are to embrace such a pluralistic methodology.

8.2 Preliminaries

Let us begin with a few preliminaries.[3] A (relativistic) *model of the universe* is an ordered pair (M, g) where M is a smooth four-dimensional "manifold" representing the shape of the universe and g is a smooth relativistic "metric" encoding the geometry of the universe. Each point in the manifold represents a possible event in space and time. Experience seems to tell us that any event (e.g., the moon landing) can be characterized by four numbers – one temporal and three spatial coordinates. Accordingly, the local structure of a manifold "looks like" a four-dimensional Cartesian coordinate system. But the global structure can be quite different. Many two-dimensional manifolds are familiar to us: the plane, the sphere, the torus, and so on.

Manifolds are good for representing events in the universe. But, the metric tells us how these events are related to one another. In particular, the "causal structure" between events is of special interest to us here. Consider a model of the universe (M, g) and a given event p in M. Which events in M can be causally influenced by p? The metric tells us. We can think of g as a kind of smooth function, which assigns lengths to all vectors at all points in M – either positive, negative, or zero. This partitions the vectors at each point into a cone structure where zero-length vectors make up the boundary of the cone while positive and negative length vectors fall, respectively, inside and outside of that boundary. Physically, vectors at a point represent velocity vectors. Since light always moves with a velocity vector of length zero, the cone structure determined by g can be thought of as demarcating the "speed of light" in all directions. Central to general relativity is the idea that "nothing can travel faster than light" and that includes any causal influences. So physically, velocity vectors must fall on or inside the boundary of the light cone. A velocity vector with this property is said to

---

[3] For details on general relativity, see Hawking and Ellis (1973), Wald (1984), and Malament (2012). For less technical introductions to global spacetime structure, see Geroch and Horowitz (1979) and Manchak (2013).

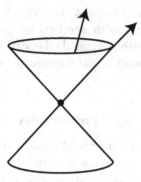

Figure 8.1 An event with associated cone structure is depicted along with a pair of
causal vectors (one spatial dimension has been suppressed).

be *causal*; if a causal vector falls strictly inside the light cone, it is *timelike*
(see Figure 8.1).

Let us follow standard practice and suppose that, ranging over the
entire manifold M, one can label the two lobes of the cone structure as
"past" and "future" in a continuous way. If a model has this property, we
say it is *time-orientable*. Physically, such a model can be given a "direction
of time" in a global sense. (A model, which is not time-orientable, can be
constructed by using a Möbius strip as the underlying manifold.) And let
us also suppose that such a labeling has been carried out at every point in
M. Now we are in a position to answer the question discussed earlier: an
event q in M can be causally influenced by event p in M only if there is
a smooth curve on M, which starts at p and ends at q, and whose tangent
vector is always (i) causal and (ii) pointed in the future direction. A curve
such as this is a *future-directed causal* curve. A future-directed causal curve
is a *future-directed timelike* curve if all of its tangent vectors are timelike.
Let us say that the collection of all points q in M such that there is
a future-directed causal curve from p to q is the *causal future* of p.
Analogous constructions yield the *causal past*, *timelike past*, and *timelike
future* of p.

Let us now consider four properties, which will form a simplified
hierarchy of causal conditions. Let (M, g) be a time-orientable model of
the universe. If there is an event p in M such that the causal past of p is all
of M, we say it has a *God point*.[4] We can think of the causal past as

---

[4] Thanks to Zvi Biener and Chris Smeenk for the terminology.

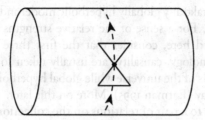

Figure 8.2 A non-chronological model of the universe is depicted with a self-intersecting future-directed timelike curve (two spatial dimensions have been suppressed).

representing the region of the universe that can possibly be observed by an idealized person at p. (After all, if q is an event located outside of the causal past of p, then how can q be observed at p if no causal influence can go from q to p?) So, a model universe has a God point if it has an event from which an idealized person can possibly observe the entire universe.

Our next two causal properties preclude "causal loops" of certain types. Let (M, g) be a time-orientable model of the universe. If the model contains a future-directed timelike curve, which intersects itself, we say it is not *chronological*. The models that fail to be chronological allow for "time travel" in the sense that an idealized person in the model can both begin and end a journey at the very same event (see Figure 8.2). A slight variant of the chronology property rules out self-intersecting future-directed causal curves; if a model satisfies this condition, then we say has the *causality* property. It is immediate that every causal model is chronological; one can show the converse does not hold.

Our final causal property ensures that a model universe is so causally "well-behaved" that it might be considered "deterministic" in some sense. Let (M, g) be any model of the universe satisfying causality. If it is the case that for any events p, q in M, the intersection of the causal past of p and the causal future of q is a "compact" region, then the model is said to be *globally hyperbolic*. A globally hyperbolic model can be split into a "stack" of three-dimensional "spatial universes" along a one-dimensional "time." In such a model, information about the structure of the spatial universe at one time can be used to determine that structure at all times. By definition, every globally hyperbolic model is causal. The converse is not true: for a counterexample, take any globally hyperbolic model and remove one point from the manifold. For a sense of the relative strengths of the four causal properties examined here, consider that the first three (non-existence of a God point, chronology, causality) are usually taken to be satisfied by all "reasonable" models of the universe while global hyperbolicity is only sometimes taken that way (Earman 1995). More on this later.

Now let us turn to a pair of relations on the collection of models of the universe. Let us say that the models (M, g) and (M', g') are *isometric* if there is a smooth one-to-one correspondence between the points in M and the points in M' which preserves all metric structure. Isometric models of the universe are physically identical. Because the isometry relation is an equivalence relation, we can partition the collection of all models into corresponding isometry equivalence classes. A similar relation is also useful: Let us say that the models (M, g) and (M', g') are *locally isometric* if (i) for every point p in M, there is a "local neighbourhood" around p, a point p' in M', and a "local neighbourhood" around p' such that the neighbourhoods are isometric and (ii) likewise with the roles of (M, g) and (M', g') interchanged. If two models of the universe are locally isometric, they share the same "local" physics. Local isometry is also an equivalence relation; indeed, the partition given by the isometry relation is just a "refinement" of the partition given by the local isometry relation; any two isometric models are locally isometric but not the other way around. Let us say that a property of a model is a *local* property if, for any two locally isometric models, one model has the property if and only if the other does as well. We say a property is a *global* property if it is not local. It turns out that each of the three causal properties considered earlier count as global. On the other hand, the "energy conditions" which limit the distribution and flow of matter can be used to define a family of local properties (see Curiel 2017).

8.3   On the Unknowability of the Universe

We are now ready to consider Feyerabend's claim that:

> (#) "the world which we want to explore is a largely unknown entity". (Feyerabend 1975/2010, p. 4).

It seems natural to interpret (#) as a type of defense for the pluralistic methodology outlined by Feyerabend; if (#) is true, we ought to "keep our options open" so that that we do not prematurely close down alternatives, which seem "unreasonable" now but which may, at some later time, point the way to some "deep-lying secrets of nature" (p. 4).[5] Even though (#) is not elaborated upon in the introduction to *Against Method*, the claim does not seem to be an off-hand remark; much of Feyerabend's historical work concerning case studies (in *Against Method* and elsewhere) serves to bolster the position that certain (e.g., empirical) methodologies would have closed off alternatives we now judge to be quite "reasonable".[6] This "keep our options open" defense of Feyerabend's methodological pluralism is distinct from the "benefits of competition" defense and, from what I understand, has yet to become a focus in the literature so far.[7]

Let us now work to show a sense in which (#) in true within the context of general relativity. First, consider that, in any model, empirical observations at an event must be confined to the casual past of that event for the reasons mentioned above. Following Clark Glymour (1977) and David Malament (1977), let us say that a model (M, g) is *observationally indistinguishable* from a model (M', g') if, for each point p in M there is a point p' in M' such that the causal pasts of p and p' are isometric (i.e., share the same physical structure). The physical significance of the definition is this: If one model is observationally indistinguishable from another model, then no idealized person represented in the first model has the epistemic resources to tell the difference between the first and second models; in other words, an idealized person in the first model cannot know that she inhabits the first model. It turns out that models can be different (i.e., non-isometric) and yet observationally indistinguishable. For example, start with the "de Sitter" model where the manifold is cylindrical and the vertical cone structures narrow ever more rapidly the more distant they are in the "past" and "future" directions. The upshot is that the causal past of any event in the model does not "wrap all the way around" the cylinder.

---

[5] Feyerabend 1975/2010, p. 4. The "keep our options open" defense of pluralism is often called the "hedging our bets" defense. For example, here is Kitcher (1993, 344): "Intuitively, a community that is prepared to hedge its bets when the situation is unclear is likely to do better than a community that moves quickly to a state of uniform opinion."

[6] Consider the following (Feyerabend 1975/2010, p. 115): "Now, what our historical examples seem to show is this: there are situations when our most liberal judgments and our most liberal rules would have eliminated a point of view which we regard today as essential for science, and would not have permitted it to prevail – and such situations occur quite frequently. The ideas survived and they now are said to be in agreement with reason. They survived because prejudice, passion, conceit, errors, sheer pigheadedness, in short because all the elements that characterize the context of discovery, opposed the dictates of reason and because these irrational elements were permitted to have their way. To express it differently: Copernicanism and other 'rational' views exist today only because reason was overruled at some time in their past."

[7] For an exception along these lines, see Oberheim (2005) on the role of incommensurability in Feyerabend's thought – especially as it relates to Duhem. For excellent discussions of the "benefits of competition" defense, see Lloyd (1997) and Bschir (2015).

Now consider a similar model where the cylinder in "unrolled" but the metric structure remains the same. The two models are observationally indistinguishable since, for any event in either model, one can find a similar event in the other model such that the two causal pasts have the same metric structure (see Figure 8.3). Note that the relation of observational indistinguishability, although reflexive and transitive, is not symmetric; there are situations where one model is observationally indistinguishable from another but not vice versa. Consider, for example, the "bottom half" of the de Sitter model – it is observationally indistinguishable from the de Sitter model but not the other way around.

What is the relation of observationally indistinguishability like? Which models of the universe have a non-isometric but observationally indistinguishable counterpart? Malament (1977) conjectured that *any* model without a God point must be related to another model in just this way. This conjecture turns out to be true but even more can be said: the result goes through even if one requires any collection of local properties to be satisfied. So even under a robust type of inductive reasoning – that the local structure of universe is the same everywhere – the underdetermination remains. Consider the following (Manchak 2009):

> *Proposition 1*: Consider a model of the universe with any collection of local properties. If the model fails to have a God point, it is observationally indistinguishable from some other (non-isometric) model with all of the same local properties.

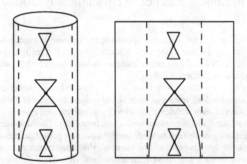

Figure 8.3 The de Sitter and "unrolled" de Sitter models are depicted, both with a representative causal past bounded by the dotted lines. Each model is observationally indistinguishable from the other (two spatial dimensions have been suppressed).

The proposition shows a sense in which any idealized person represented in virtually any model will not have the epistemic resources to know which model she inhabits. For her, the universe is (and will always remain) a largely unknown entity. But how serious is this epistemic predicament? Perhaps it is the case that, although the full structure of a model universe is unknowable from within, partial knowledge can be obtained – again, with a robust type of inductive reasoning – concerning some global properties of interest. It turns out even this cannot be done for many important global properties (e.g., global hyperbolicity) thought to be satisfied by some "reasonable" models of the universe. Here, we will focus attention on one property in particular: inextendibility.[8] Roughly, this property requires that a model universe be "as large as it can be" in a natural sense. We say a model of the universe (M, g) is *extendible* if there is another model (M', g') such that M is isometric to a proper subset of M'; here (M', g') is a (proper) *extension* of (M, g). A model (M, g) is *inextendible* if it is not extendible (see Figure 8.4).

The idea that all "reasonable" models of the universe must be inextendible is more or less taken for granted within the community. The reasoning behind this position is summarized by John Earman (1995, pp. 32–33):

> Metaphysical considerations suggest that to be a serious candidate for describing actuality, a spacetime should be [inextendible]. For example, for the Creative Force to actualize a proper subpart of a larger spacetime would seem to be a violation of Leibniz's principles of sufficient reason and

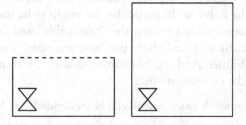

Figure 8.4 The first model is extendible. The second model is an extension of the first and is inextendible (two spatial dimensions have been suppressed).

---

[8] See Geroch (1970) and Clarke (1976) for details concerning this property. The appendix of Geroch (1970) is an especially good resource; it contains a list of precise foundational questions concerning inextendibility – many of which remain open.

plenitude. If one adopts the image of spacetime as being generated or built up as time passes then the dynamical version of the principle of sufficient reason would ask why the Creative Force would stop building if it is possible to continue ... Some readers may be shocked by the introduction of metaphysical considerations in the hardest of the "hard sciences." But in fact leading workers in relativistic gravitation, though they don't invoke the name of Leibniz, are motivated by such principles.

We will return to the metaphysical justification of inextendibility in the next portion of the paper. For now, let us turn to the strengthened underdetermination result mentioned above (Manchak 2011).

> *Proposition 2*: Consider a model of the universe with any collection of local properties. If the model fails to have a God point, it is observationally indistinguishable from some other (non-isometric) model with all of the same local properties which is also extendible.

The proposition shows a sense in which any idealized person in virtually any model cannot know – even with a robust type of inductive reasoning – that her universe is inextendible. And since inextendibility is taken to be a necessary property of all "reasonable" models, the upshot is this: within general relativity, there is (and will always remain) the epistemic possibility that the universe is best represented by an "unreasonable" model.

8.4 On "Unreasonable" Universes

The underdetermination results just presented seem to lead naturally to the position that we ought to "keep our options open" with respect to competing models of the universe. In particular, we ought to be suspicious about various distinctions made between the "reasonable" and "unreasonable" models – especially in cases where the "unreasonable" ones remain live epistemic possibilities. And yet, as mentioned above, it seems to be taken for granted by the community that

(†) "any reasonable space-time should be inextendible". (Clarke 1993, p. 8).

Just how entrenched is the (†) position? Like inextendibility, there are a number of other properties which are often considered necessary for all "reasonable" models of the universe; as mentioned earlier, chronology is one such. But research is still routinely conducted on models that fail to have many of these properties – consider the vast and flourishing literature on "time travel" and "time machines," for example (Earman et al. 2016). It is of some interest that this is not the case with respect to the failure of inextendibility; the subject has given rise to almost no literature at all.[9] Apparently, the negation of (†) is an especially "unreasonable" position. Still, I do not think it has been "given all the chances it deserves" (Feyerabend 1975/2010, p. 29).

In what follows, I will proceed according to the counter-inductive suggestion to "introduce and elaborate hypotheses which are inconsistent with well-established theories" (p. 13). In particular, I will introduce and elaborate the negation of (†). Why work to proliferate an especially "unreasonable" position in this way? One reason has already been emphasized: if (†) is wrong, we do not want to prematurely settle on it. Another reason is this: even if (†) were "right" in some sense, the development of its negation via a "process of competition" (p. 14) only serves to improve our understanding of the (†) position itself. One is guided by the "hope that working without the rule, or on the basis of a contrary rule we shall eventually find a new form of rationality" (Feyerabend 1977, p. 368). It is this prospect of a "new form of rationality" which amounts to an additional defense of Feyerabend's methodological pluralism (see Shaw 2017). As we will soon see, this prospect is realized in the present case. As a result of elaborating on the negation of (†), the usual lines between "reasonable" and "unreasonable" models of the universe will seem to blur. In addition, entirely new ways of looking at the situation will present themselves. The entire undertaking proceeds in the spirit of Feyerabend as one might expect: one "plays the game of Reason in order to undercut the authority of Reason" (Feyerabend 1975/2010, p. 16).

We begin by drawing a distinction between the "reasonable" and "unreasonable" models of the universe; let $\mathcal{U}$ be the collection of all models and let $\mathcal{R} \subset \mathcal{U}$ be a working collection of "reasonable" ones. Such a collection proves useful to consider even if we have yet to pin down its makeup. It is crucial in what follows that that we do not suppose from the outset that that every model in $\mathcal{R}$ is inextendible since this is the very question under investigation. But this opens the way for a curious possibility: perhaps there is an "extendible" model of the universe in $\mathcal{R}$ which nonetheless cannot be "reasonably" extended in the sense that all of its extensions fail to be in $\mathcal{R}$. Would not such a model be "as large as it can be" in the only sense that mattered? Would it not be "reasonable" to demand that such a model be called "inextendible"?

---

[9] Indeed, work has shifted somewhat dramatically toward an exploration of even stronger definitions of inextendibility; an "inextendible" model can sometimes be extended if the metric is not required to be smooth. See Galloway and Ling (2017) and Sbierski (2018) for the latest twists and turns.

It turns out that scenarios like the one just mentioned not only can be constructed – but also they arise quite naturally when various "reasonable" hypotheses are entertained. For example, consider the so-called (strong) cosmic censorship conjecture of Roger Penrose (1979). The content of the conjecture can be expressed as the position that

(††) all "reasonable spacetimes are globally hyperbolic". (Wald 1984, p. 304).

It should be noted that, unlike (†), the position (††) is not uncontroversial (Earman 1995). But, whatever else is the case, it is not "unreasonable" to consider its consequences. Suppose (††) is true; suppose the collection $\mathcal{G}$ of globally hyperbolic models is such that $\mathcal{R} \subset \mathcal{G}$. Now consider the "Misner" model, where the manifold is cylindrical and the cone structures "tip over" as they go up the cylinder (see Figure 8.5). The "bottom half" of the Misner model – call it the "lower Misner" model – counts as extendible when taken as a model in its own right (the Misner model itself being just one of its many extensions). But aside from its extendibility, the lower Misner model checks all of the usual boxes required of all "reasonable" models. In particular, it is both globally hyperbolic and a "vacuum solution" of Einstein's equation. This latter fact ensures that the model satisfies all of the energy conditions mentioned earlier, which guarantee a "reasonable" distribution and flow of matter. Indeed, the lower Misner model is about as "reasonable" as an extendible spacetime can be. Given that we have deliberately left open the possibility that $\mathcal{R}$ contain extendible models, it seems "reasonable" to consider the case where the lower Misner model is found in that collection. But now observe: one can show that every extension of the lower Misner model fails to be in $\mathcal{G}$ (Manchak 2017). By (††), every such extension fails to be in $\mathcal{R}$ as well. The situation seems to be this: we have a "reasonable" model of the universe, which

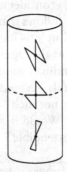

Figure 8.5 The Misner model. The region below the dotted line is a globally hyperbolic vacuum solution when taken as a model in its own right(two spatial dimensions have been suppressed).

cannot be "reasonably" extended and yet counts as "extendible" according to the standard definition.

Due to examples like the one just given, Bob Geroch has considered the possibility of revising the standard definition of inextendibility. For example, a variant definition can be constructed for each property $\mathcal{P} \subset \mathcal{U}$: let us say that a model is a $\mathcal{P}$-*model* if it is in the collection $\mathcal{P}$ a $\mathcal{P}$-model is $\mathcal{P}$-*extendible* if it has an extension in $\mathcal{P}$ – such an extension is a $\mathcal{P}$-*extension* – and $\mathcal{P}$-*inextendible* otherwise. But such "unpleasant modifications" to the standard definition of inextendibility seem to be unnecessary if, for a variety of "reasonable" properties $\mathcal{P} \subset \mathcal{U}$ the following is true (Geroch 1970, 278).

(*) Every $\mathcal{P}$-inextendible $\mathcal{P}$-model is inextendible.

Let us try out some properties $\mathcal{P}$ in an attempt to get a grip on the situation. It is immediate from the lower Misner example that (*) is false if $\mathcal{P}$ is the collection $\mathcal{G}$ of globally hyperbolic models. What about other "reasonable" properties of interest? A simple example shows (*) to be false if $\mathcal{P}$ is the collection $\mathcal{C}$ of models which satisfy causality. And, we have recently learned that (*) is false if $\mathcal{P}$ is the collection $\mathcal{E}$ of models satisfying the "weak" energy condition (Manchak 2017). More work is certainly needed. Still, at present no significant "reasonable" property $\mathcal{P}$ has been found which renders (*) true. It seems "reasonable" to explore revisions to the standard definition of inextendibility.

Let us take a step back. Recall that the primary justification for (†) seemed to rest on Leibniz's principles of sufficient reason and plenitude. Here is a representative statement along these lines (Geroch 1970, p. 262): "Why, after all, would Nature stop building our universe … when She could just as well have carried on?" Underpinning the metaphysical views expressed by Geroch and others is this central fact (Geroch 1970).[10]

Proposition 3: Every extendible model has an inextendible extension

To be sure, the proposition is beautiful in its simplicity and power. But given the need to consider revisions to the definition of inextendibility, is it not "reasonable" to investigate whether the following is true for a variety of "reasonable" properties $\mathcal{P} \subset \mathcal{U}$?

---

[10] It is of some interest that this proposition is one of the few results in general relativity which seems to depend crucially on the axiom of choice for its proof. See Clarke (1976).

Table 8.1 *A synopsis of the situation so far*

	$\mathcal{U}$	$\mathcal{K}$	C	$\mathcal{G}$	$\mathcal{E}$	$\mathcal{B}$	...
(*)	T	?	F	F	F	?	...
(**)	T	T	?	?	?	F	...

(**) Every $\mathcal{P}$-extendible $\mathcal{P}$-model has a $\mathcal{P}$-inextendible $\mathcal{P}$-extension.

Very little is known concerning the status of (**) with respect to "reasonable" properties of interest.[11] We do know that (**) is true if $\mathcal{P}$ is the collection $\mathcal{K}$ of chronological models (Manchak 2017). Perhaps this gives comfort to those wishing to defend (†). But, there are also examples which also go the other way. Let us say that a model has the *big bang* property if every "straight" causal curve in the model ends in a "singularity" in the past direction. It turns out that (**) is false if $\mathcal{P}$ is the collection $\mathcal{B}$ of big bang models (Manchak 2016). We see that under some "reasonable" revisions to the definition of inextendibility, it is not always possible for a "reasonable" model of the universe to be "as large as it can be." Reason seems to have led us here: perhaps (†) is wrong. (See Table 8.1 for a synopsis of the situation so far.)

8.5 Final Remark

By proceeding counter-inductively with respect to (†), we have stumbled upon an even more fundamental way to "introduce and elaborate hypotheses which are inconsistent with well-established theories" (Feyerabend 1975/ 2010, 13). Let me explain. For a number of "reasonable" properties $\mathcal{P} \subset \mathcal{U}$, the following position concerning general relativity seems natural.

> One might now modify general relativity as follows: the new theory is to be general relativity, but with the additional condition that only [$\mathcal{P}$] space-times are permitted. (Geroch 1977, p. 87)

So for each property $\mathcal{P}$, we have a variant theory of general relativity – call it GR($\mathcal{P}$). At once we find ourselves swimming in an "ocean of mutually incompatible alternatives" (Feyerabend 1975/2010, p. 14). We know that the property of inextendibility "works differently" in some of these variant

---

[11] Presumably, the axiom of choice (in the form of Zorn's lemma) can be applied straightforwardly to obtain positive results for some "reasonable" properties (e.g., the energy conditions). But this route will not work in other cases (Low 2012).

Table 8.2 *Results concerning variant theories of general relativity*

	GR($\mathcal{U}$)	GR($\mathcal{K}$)	GR($\mathcal{C}$)	GR($\mathcal{G}$)	GR($\mathcal{E}$)	GR($\mathcal{B}$)	...
(⋆)	T	T	?	?	?	F	...
⋮	⋮	⋮	⋮	⋮	⋮	⋮	

theories than it does in the standard one; if (*) is false for some $\mathcal{P} \subset \mathcal{U}$, then there is a model universe in $\mathcal{P}$, which is "inextendible" according to GR($\mathcal{P}$) but "extendible" according to GR($\mathcal{U}$). So, a question like "is the model inextendible?" can be interpreted in as many ways as there are variant theories permitting the model under consideration. Given the state of affairs, a study of the property of inextendibility from within each alternative would seem to be quite appropriate. From the work mentioned earlier, we find that foundational claims like the following come out as true in some variant theories and false in other theories.

(⋆) Every extendible model has an inextendible extension.

Other statements of interest could be explored in this way. For example, some precise statements concerning "singularities" and "time machines" depend on the definition of inextendibility (Geroch 1970; Earman et al. 2016). Statements concerning the "stability" of various properties (e.g., causal properties) could also be investigated (cf. Hawking 1969). A table presents itself in which such statements are the rows and the columns are variant theories of general relativity (see Table 8.2).

By contrasting one variant theory with another, we come to understand "general relativity" in a more nuanced way. In particular, this pluralistic methodology awakens us to the fact that positions like (†) cannot possibly be settled – if they are to be settled at all – before the requisite work is completed. Just imagine the number of question marks implicit in Table 8.2![12,13]

[12] Even if one restricts attention only to "reasonable" properties consisting of combinations of standard causal and energy conditions, dozens of "reasonable" variant theories of general relativity can be easily constructed. If non-standard causal conditions are permitted, the number of "reasonable" variant theories becomes infinite (see Carter 1971). This is to say nothing of the "unreasonable" alternatives.

[13] Acknowledgments: My thanks to Jeff Barrett, Thomas Barrett, Karim Bschir, Erik Curiel, Amanda Knox, Martin Lesourd, David Malament, Meka Manchak, Helen Meskhidze, Haylee Millikan, Jamie Shaw, Chris Smeenk, Kyle Stanford, Jim Weatherall, Nathan Westbrook, and Chris Wüthrich for helpful discussions and/or comments on a previous version of the paper.

CHAPTER 9

Feyerabend, Science and Scientism

Ian James Kidd

9.1 Introduction

Paul Feyerabend (1924–1994) acquired a variety of epithets during the latter stage of his career. The most persistent is perhaps *Nature*'s description of him as 'the worst enemy of science', later adopted as the title of an important collection of essays in his honour. The label encapsulates his bad reputation, at least within philosophy of science, which can be divided roughly into two aspects.

The first are criticisms of the actual or perceived content of his work, most usually that he was, at least at certain points in his career, anti-science, pro-pseudoscience and a radical relativist or perhaps post-modernist. Some of these can be easily rebutted. Considered closely, the putative 'defences' of astrology, parapsychology, witchcraft and alternative medicine turn out to be nothing of the sort (Kidd 2013, 2016a, 2018). His general strategy was to point out the epistemic failings of those scientists who dismissed such beliefs and practices without any properly informed understanding of them. Astrology was 'bunk', he argued, but one needs better arguments against it than those typically offered by those whose social authority owes to their elevated epistemic standing. In the case of relativism and other alleged philosophical sins, recent work by Martin Kusch (2016) and Lisa Heller (2016) tell a more complex story: the ultra-relativism of *Science in a Free Society* modulated, slowly, into those thirteen 'relativistic theses' in *Farewell to Reason*, most then rejected by the time of *Conquest of Abundance*. Similarly, there are no good reasons to regard him as a post-modernist, at least on three substantive characterisations of that capacious term (Kidd 2016b). As to the other charge – that Feyerabend was 'anti-science' – refuting that is the aim of this chapter.

The second aspect of Feyerabend's bad reputation is less easily disposed of, since it is rooted in criticisms of his professional conduct. Certainly, polemic, rhetoric and a jocular tone are not to everyone's taste, but there are other, more serious complaints. Consider the rudeness and wilful vagary that often crept into his writing, the most egregious instance of which would surely be the third section of *Science in a Free Society* – the fieriest of his books – which reply to critics of *Against Method* and was named, by Feyerabend, 'Conversations with Illiterates'. Eric Oberheim opens his book, *Feyerabend's Philosophy*, with honest documentation of the rhetorical and provocative excesses, dubious self-testimonies and scathing tone and content of much of Feyerabend's work, at least during the over-heated writings of the late 1970s.

Since such failings are not in doubt, my aim is not to defend every-thing that he said, nor the ways that he said them. My stance on his work is one of critical sympathy, respecting both the principle of charity and the fact that Feyerabend often put considerable pressure upon it. We need to edit, amend and augment Feyerabend's ideas, alert to his rhetorical and scholarly failings and their consequent interpretive problems, for many of which he was culpable. What we end up with is something that merits the label 'Feyerabendian', taking the best of what he offers, and refining or removing the weaker material. An important core theme of what we end up with is an enduring *anti-scientism*.

9.2   Critique

From the late 1960s onwards, an increasingly central feature of Feyerabend's work was experiments with different ways of motivating and performing critical scrutiny of prevailing conceptions of science. By that latter term, I mean broad accounts of the nature, scope and value of science, such as Thomas Kuhn's account of paradigms, Karl Popper's falsificationism, Imre Lakatos's methodology of scientific research programmes and other more modest efforts by less-esteemed *Theoretiker der Wissenschaft*. An important role for philosophy of science is to create such accounts, but also to subject them to critical analysis.

An unusually clear statement of Feyerabend's conception of the aims of philosophy of science as a critical discipline is a 1976 paper, 'On the Critique of Scientific Reason.' Its title is a play on Kant's *Kritik der reinen Vernuft*, where 'critique' refers to a systematic investigation of something's nature and limits, such as reason, for Kant. What Onora O'Neill says of Kant's critique of reason is also true of Feyerabend's critique of scientific reason as follows:

> Whatever else a critique of reason attempts, it must surely criticise reason. Further, if it is not to point toward nihilism, a critique of reason cannot have only a negative or destructive outcome, but must vindicate at least some standards or principles as authorities on which thinking and doing may rely, and by which they may (in part) be judged. (O'Neill 1992, p. 280)

Swap 'reason' for 'science', in this passage, and one has a reasonable account of Feyerabend's vision of the spirit of critical philosophy of science. Naturally, his execution of that critique tended to excess, at least in the more polemical writings, such as *Science in a Free Society*. But, the project itself is a good one, which Feyerabend characterised using the following 'fundamental questions': There are two questions that arise in the course of any critique of scientific reason. They are as follows:

(a) *What is science* – how does it proceed, what are its results, how do its procedures, standards and results differ from the procedures, standards and results of other enterprises?
(b) *What's so great about science* – what makes sciences preferable to other forms of life, using different standards and getting different kinds of results as a consequence? (Feyerabend 1976, p. 110).

A critique of scientific reason therefore challenges our conceptions of the historical development, aims and values, methods and practices, theoretical and ontological commitments, and social organisation of the motley of activities gathered under the capacious label 'science'. Later in the paper, Feyerabend complains that this 'critical attitude is only rarely found among [contemporary] philosophers of science' (p. 112) – a claim to which I return later.

Starting from those two 'fundamental questions', I suggest that this critique devolves into two sub-critiques, each evident in Feyerabend's own writings. First, a *critique of science's self-understanding*, motivated by the curt, concise question, *what is science?*, which occupied Feyerabend from at least the mid-1960s. It would have been familiar to mid-twentieth century analytic philosophers of science, focussed as many of them were on either expanding inherited logical positivist conceptions of science, or proposing their own alternatives. Second, a *critique of scientific modernity*, directed at the problems and anxieties – moral, socio-political, even spiritual – emergent within 'forms of life' shaped by scientific knowledge, institutions and *Weltanschauung*. It was increasingly evident from the mid-1970s, reflecting an aspiration to cultural critique that owed as much to Feyerabend's Viennese roots and temperament as to the social and political radicalism of the period.

Certainly, there were few allied voices among philosophers of science of the time. An exception was Jerry Ravetz, whose 1971 book, *Science and its Social Problems*, focusses on the connections between 'the problems of the character of scientific knowledge, of the sociology and ethics of science, and of the applications of science to technology and to human welfare' (Ravetz 1971, p. 1). The merging of critiques of science's self-understanding and wider features of social and political culture are obvious. Such themes were soon to become central to the emerging movements of feminist and post-colonial science studies and were, of course, already resonant in the wider tradition of continental philosophy (Gutting 2005).

In this chapter, my aim is to sketch the programmatic structure of the critique of scientific reason as developed by Feyerabend throughout his writings from roughly the early 1970s, to the very late writings, up to his death in 1994. The two sub-critiques should be understood as experiments in anti-scientism, as efforts to motivate and enact critical investigations of the nature, scope and value of science – to free us from various 'myths' about science that, if left unchecked, tend to spread distorted ideas about science that risk leading us into 'dogmatism', 'tyranny' and other epistemic sins.

Although, to my knowledge, Feyerabend never used the term 'scientism', it usefully gathers the themes and tone of many of his remarks on attitudes towards and ideas about science within both academia and wider culture. Scientism, broadly stated, concerns a distorted or exaggerated estimate of the nature, scope and value of science, and so can take many different forms, as mapped out by Mikael Stenmark (2001) and Rik Peels (2018). In the case of Feyerabend's critique of scientific reason, the worries about scientism mainly emerge from a failure by philosophers of science to subject the sciences to critical scrutiny. Among the many reasons for that failure, an important one is certain contingent developments in the discipline of philosophy of science.

Excellent historical work by Heather Douglas (2009) and George Reisch (2005) has revealed the important influence on the concerns and agenda of philosophy of science in the United States of the culture and politics of the cold war, which in turn was part of what Feyerabend was reacting against – or so I have argued (Kidd 2016c). Confining attention to abstract issues of logic, methodology and epistemology may have been important to the survival of the discipline during the 1950s, to protect a fledgling discipline caught in a storm of political paranoia, when universities were reconceptualised as the intellectual frontlines against the Soviets. Since science was a source of technologies, knowledge and prestige essential to the ideological and military aspects of the Cold War, it was pragmatic for philosophy of science to adopt the role of cheerleader, while protecting itself from political scrutiny using the value-free ideal.

By the early 1970s, however, the constrained agenda of philosophy of science was frustrating Feyerabend. What had been expedient twenty years earlier was now becoming myopic, distracting attention away from increasingly urgent questions about the epistemic, practical and cultural effects of science. In a neglected 1968 paper, 'Science, Freedom, and the Good Life', he worries that the science is failing to 'serve the increase of human happiness', due to the 'disjointed' character of the 'various departments' of our 'intellectual enterprise', which Feyerabend thought could be repaired, if we 'eliminate the dogmatism of contemporary scientific inquiry' (Feyerabend 1968b, pp. 134ff). Philosophers of science bore a double fault, here: first, for contributing to the 'disjointed', 'dogmatic' character of science by providing conceptions of science premised on a sharp demarcation between science and non-science; second, for celebrating dogmatism as an integral feature of effective scientific enquiry – the targets here being Popper and Kuhn, respectively. Changing political and cultural conditions requires a reappraisal of the agenda and character of philosophy of science, ideally in the direction of a robust critique of scientific reason.

Unfortunately, the sensible case for a more critical and socially engaged philosophy of science was occluded by Feyerabend's failures to present it as an attractive new programme for the discipline. In such papers as 'Philosophy of Science: A Subject with a Great Past' and '*Die Wissenschaftstheorie – eine bisher unbekannte Form des Irrsinns?*', the tone and claims are derogatory rather than diplomatic (and, in the latter, also insultingly ableist). Even the most earnest of Feyerabend's admirers should concede that he lacked the professional tact and vision that is necessary to properly effect the sort of disciplinary reconceptualisation to which he aspired. What follows, then, is an attempt to offer a soberer account of the critique of scientific reason, inspired by and critically respectful of Feyerabend's remarks.

I judge that the critique of science's self-understanding is the more sophisticated and successful than the critique of scientific modernity, since the latter suffers from its lack of engagement with various relevant philosophical traditions. Each critique offers various ways of conceiving and criticising forms of scientism, consistent with Feyerabend's characteristic intellectual practice of being actively pluralistic and experimental, not in dogged pursuit of a single idea. To a degree, this tendency explains the deficiencies of the critique of scientific modernity – doggedness is, after all, not the same as dogmatism, and a patient, plodding style might not have suited Feyerabend, even if it could have yielded a more consistently successful project in philosophical anti-scientism.

Appraisal of that claim, though, requires a better understanding of the content of the two sub-critiques of scientific reason, to which I now turn.

9.3 Critique of Science's Self-Understanding

At its most general, a critique of science's self-understanding aims to interrogate prevailing conceptions of the historical development, methods, practices, values, aims and social and theoretical structures of the sciences. Some critiques challenge existing conceptions, others proffer their own, while many do both at the same time. The construction of conceptions of science was very much the main business of mainstream philosophy of science of the middle half of the last century, exemplified by Popper, Kuhn, Lakatos and their distinguished predecessor, logical positivism.

Such conceptions tend to be more or less schematic, offering broad pictures of the structure and processes of science, albeit typically focussed onto the physical sciences. They are also prescriptive as well as descriptive, offering normative visions of how science ought to be organised, rather than an account of what has come to be its organisation. An important part of Feyerabend's critique of Kuhn, recall, was an 'ambiguity of presentation', concerning the issue of whether the account of science set out in *Structure of Scientific Revolutions* was prescriptive or normative (Feyerabend 1970a, §2) – the brisk reply to which was that it was both (see Kuhn 1970, p. 237). But there are three other features of conceptions of science that are more centrally related to a critique of science's self-understanding.

To start with, they are integral to our efforts to understand the place of science within our 'form of life', at least some of the time. Certain practical or policy questions may proceed at the level of some specific theory, such as the theory of evolution, for sure, but sometimes the concern is that larger, more amorphous thing, 'science'. Second, conceptions of science are typically inherited and accepted without due critical diligence, usually through the processes of education and socialisation. It's rare that they are subjected to processes of informed deliberation and decision, such that those who employ them have a reflective relationship to them.

In itself, this might not be problematic, but – and this is the third feature – there is, at least to Feyerabend's mind, a constant danger that these conceptions may become objects of uncritical acceptance. Complacency, dogmatism and other dodgy epistemic attitudes can cause real dangers when their object is something as socially and epistemically powerful at the sciences. Hence, the rather apocalyptic tone of the warning that Kuhn's model of science was 'bound to tend to increase the anti-humanitarian tendencies' of 'much of post-Newtonian science' (1970, p. 197–198) and that 'any method that encourages uniformity ... enforces an enlightened conformism [and] leads to a deterioration of intellectual capabilities' (Feyerabend 1975a p. 45). By opposing such models of science, one was actually pursuing a humanitarian agenda, hence Feyerabend's explanation that the motivation to write *Against Method* was 'humanitarian, not intellectual' (Feyerabend 1975/1993, p. 3).

Against Method was actually the title of four separate works, a very long essay of 1970 and the three editions of the book authored by Feyerabend, which were intended as a dialogue to be conducted with Imre Lakatos. True to their author's description of them as 'collages', there is overlap and copying and pasting across these versions, such that each differs from the other in content, often in quite substantive ways. Moreover, such interpretive complexities are further muddied by the well-known problems of Feyerabend's idiosyncratic style and referencing (see Oberheim 2006, Chapter 1).

We can, though, arguably identify a stable core to these works, centred on their various challenges to *methodological monism*: a conception of science as a unitary enterprise, whose epistemic efficacy and unity owe to a common employment, across its constituent disciplines and projects, of the singular, formalised set of well defined, historically invariant, context insensitive methodological norms. I think the clearest statement of the general thesis may be the introduction to the Chinese edition:

> The thesis [of AM is]: *the events, procedures and results that constitute the sciences have no common structure*; there are no elements that occur in every scientific investigation but are missing elsewhere ... procedures that paid off in the past may create havoc when imposed on the future. Successful research does not obey general standards; it relies now on one trick, now on another; the moves that advance it and the standards that define what counts as an advance are not always known to the movers. (Feyerabend 1975/1993, p. 1, original emphasis)

Feyerabend's aim was to challenge methodologically monistic conceptions of science and replace them with more pluralistic ones – hence the colourfully named thesis, 'epistemological anarchism', a form of active normative epistemic pluralism (see Chang 2012, Chapter 5).

Feyerabend offers four main strategies of criticism against forms of methodological monism: the first two charging it with abstraction and obstruction; and the latter two targeted at its tendency to interfere with our capacities to organise and appreciate the sciences. Note, from this point, a switch from the singular, 'science', to the plural, 'sciences', consistent with Feyerabend's constant emphasis on its methodological, theoretical, ontological and axiological plurality. In some of the later work, there is still talk of science in the singular, but almost always to refer to 'a mythical, monster "science"' (Feyerabend 1975/1993, p. 245).

The first line of criticism is that the doctrines of methodological monism are detached from the history and practice of the sciences, as disclosed by the results of historical and sociological enquiry. Such unitary models of science may work nicely on paper but bear little resemblance to what actually goes on in the laboratory, at the workbench or out in the field. The famous Galileo case study used as the basis for *Against Method*, for instance, aimed to show that actual scientific activity is far more complex than the tidy reconstructions offered by the philosophers of science. The rules are often ignored without any epistemic loss, while violating them often pays dividends, meaning that there is far more scope for creative epistemic agency than is sometimes supposed – a sensible point obscured by that unfortunate provocative slogan, 'Anything goes' (see Oberheim 2006, p. 15 fn.30; Shaw 2017). If we are attentive to history and practice, we find that 'the actual sciences, as practised by scientists, have little to do with the monolithic monster "science"' (Feyerabend 1987, p. 155). As late as *Conquest of Abundance*, there are reminders that 'talk of a uniform 'scientific view of the world' can have its uses – to inspire scientists, for instance – just as long as those who indulge in such talk are alert to 'the complexities of research', without which they are apt to fall victim to 'simpleminded and most vapid tales' (Feyerabend 1999, p. 160).

A second strategy of criticism involves the claim that methodological monism would tend to disrupt the practice of science by promoting dogmas that will (a) often fail to enable and (b) often obstruct scientific enquiry. Feyerabend sought to show that following rules of method often impedes epistemic progress, which in other cases can only be achieved by the strategic suspension or violation of those rules. Method often gets us ahead, but not always, which is something that should be known to wise, experienced scientists, unless their minds are confused by the 'fairytales' and 'myths' circulated by the philosophers of science! The third critical strategy is closely related, as it concerns the social epistemological worry that monism tends to distort our ability to understand and organise the relations between sciences and other epistemic activities and resources. If our thinking about science becomes myth-ridden, then it becomes distorted by various errors, confusions and simplifications. An important consequence is confusion about the relationship between the sciences and other forms of enquiry, which is a point that is credited to Feyerabend by John Dupré:

> In general, I can imagine no reason why a ranking of projects of inquiry in terms of a plausible set of epistemic virtues (let alone epistemic and social virtues) would end up with most of the traditional sciences gathered at the top. No sharp distinction between science and lesser forms of knowledge production can survive this reconception of epistemic merit. It might fairly be said, if paradoxically, that with the disunity of science comes a kind of unity of knowledge. (Dupré, 1993, p. 243)

Belief in a single thing – the Scientific Method – that sets up a boundary between science and non-science that is too neat and tidy to capture the messy relationships that actually obtain, often fluidly, between the plurality of our epistemic projects.

While these three critical strategies become familiar enough to readers of Feyerabend, there is another, too. His standard vocabulary for criticising scientistic 'myths' is that of their being 'simpleminded' and 'crude' – that, for instance, 'received versions' of science are 'most of the time ... not only incorrect, but ... much more simpleminded' than the actual complexity of practice (*Philosophical Papers* vol. 2, p. 217), and that any 'theory of science that devises standards and structural elements for *all* scientific activities' is, inevitably, 'too crude an instrument' (Feyerabend 1975/1993, p. 1). 'Uniformity', of any sort, 'reduces our joys' (Feyerabend 1987, p. 1). Similar complaints are levelled at homogenising conceptions of creativity, which deny the 'efficiency ... modesty and, above all, the humanity of [scientific] practitioners' (1987b, p. 711). Underlying such remarks, I think, is a concern that distorting 'myths' about science are also problematic because they diminish our appreciation of the creativity, complexity and imaginativeness of scientists and scientific enquiry. Aesthetic appreciation of the imaginativeness and variety of actual scientific activity is the price paid by those who, 'undisturbed by the complexities of research, are liable to fall for the most simpleminded and most vapid tale' (Feyerabend 1999, p. 160).

A common aim of these interrelated strategies of criticism is to challenge a certain range of conceptions of science, namely, those which maintain that its epistemic efficacy and unity owe to the Scientific Method. Read this way, the various works called *Against Method* are critiquing a set of *conceptions of science*, not rejecting or derogating science itself – and not only since, for

Feyerabend, the arch-pluralist, there is no singular conception of science. *Against Method* is really against certain scientistic conceptions of scientific methods, those of a sort that distort their historical development and obscuring their complexity and creativity. The upshot is that nothing is lost by abandoning monistic visions of science, even if what one has in their place are messier and more pluralistic, such that one cannot easily justify the epistemic and social authority of science by appeal to its special methodological credentials (Feyerabend 1975a, p. 217). But rejection of one set of credentials does not entail a rejection of the authority of the sciences *tout court*, even if Feyerabend too often obscured that point with dramatic rejection of the 'excellence' of science and its separation from the state (*Science in a Free Society*, Part II, §§8–10).

I propose that we interpret the critique of methodological monism as a critique of a specific self-understanding of the sciences as a methodologically unified enterprise. Admirers of science can reject such monistic conceptions without needing to drift into extreme talk of rejecting its epistemic excellence and stripping away its privileged social authority and cultural status. One simply has to tell more complex stories about the development, structure and operations of those sciences, to provide sophisticated answers to the question, *What is science?* Any failure to take up that task represented a dereliction of the duties of philosophy of science, according to Feyerabend's normative disciplinary vision. Moreover, subsequent developments within philosophy of science have arguably confirmed to a striking degree to broad details of the messier conception of science sketched by Feyerabend.

We can clearly see this in a different emphasis of historical and sociological 'turns', the emphasis on the plurality and disunity of science of the Stanford School, and the more recent turns to values and socially relevant philosophy of science – all of which affirm Feyerabend's vision of 'science' as an historically contingent, pluralistic, disunified multitude of activities and projects, charged with values and complexly interwoven with wider social and political contexts. Interestingly, few contemporary philosophers perceive Feyerabend's prescience, even among the modern advocates of pluralism, the two honourable exceptions being John Dupré (1993, pp. 262ff) and Hasok Chang (2012, Chapter 5).

9.4 Critique of Scientific Modernity

A critique of scientific modernity aims at critical investigation of the practical, moral, cultural and existential problems that can arise when scientific ambitions, institutions and *Weltbilder* come to be entrenched within a certain culture or form of life. Such critiques need not presuppose that cultures deeply shaped by the sciences must necessarily be replete with such problems, they generally share a sense that such problems are, as a matter of fact, evident within the forms of scientific culture that have emerged in late modern societies.

Such concerns are captured by Feyerabend's pithy question, *What's so great about science?*, which devolves into various sub-questions – 'great' in what sense? and for whom? and at what cost? – and start to become prominent in his work from the early 1970s. *Science in a Free Society* directs its polemics against a homogenous bête noire, 'Western science', and *Farewell to Reason* opens with an ominous warnings of 'powerful traditions' that 'oppose' a sense of the value of 'cultural diversity', destroying the ecological and epistemic richness of the world, making life barren and meaningless' (Feyerabend 1987, p. 1). Into the very late writings, such concerns are supercharged into the titular theme of *Conquest of Abundance*, an energetic denunciation of what that books evocative subtitle calls 'a tale of abstraction versus the richness of Being'. Although that book was unfinished at the time of Feyerabend's death, its guiding theme was profound concern about powerful tendencies towards a stultifying epistemic and cultural homogeneity.

Such ventures into cultural critique are stirring and account for much of the broad interest in Feyerabend's work outside of philosophy. As John Preston reminds us, *Against Method* and *Science in a Free Society* have both garnered 'an audience far wider than books in the philosophy of science usually have', since they offered, 'a critique of the position of science within Western societies', that is 'not to be confused with a critique of science itself' (Preston 1997a, p. 5). Unfortunately, the critique was unsystematically developed, shifting in its targets and styles of criticism, often in unhelpfully vague ways.

We can organise the most consistent concerns and themes of the critique of scientific modernity under three labels. To start with, there is the charge that the sciences motivate, justify and enable projects of profound epistemic, environmental and cultural violence, which feed a 'massive trend towards natural, social, and technological uniformity' (Feyerabend 1987, p. 3). Consider, for instance, the denunciations of the erosion of global cultural and epistemic diversity at the hands of the self-serving ideologies of 'rationalism' concocted by philosophers; the attacks, the forcible displacement and subordination of aboriginal peoples, especially if disguised with a treacly rhetoric of 'progress' and 'development'; or the systematic destruction of natural environments caused by the relentless instrumentalisation of creatures and environments.

A second aspect of the critique of scientific modernity concerns the increasing prominence in social and political culture of forms of *philistine scientism*, evident in depreciation of the significance of the arts and of humanistic values, activities and sensibilities, especially in comparison with 'useful' subjects like science, engineering and technology. Within Feyerabend's writings, this concern took at least two main forms. In the late 1960s, the arts were primarily epistemic: the theatre, drama and history could provide important forms of criticism and insight, not least in relation to the sciences themselves. Feyerabend's own demonstrations of his involved careful studies of the playwrights Eugène Ionescu and Bertolt Brecht, not least the latter's *Life of Galileo* (see Feyerabend 1967a, 1967b, 1975b).

An appreciation of the critical, epistemic value of the arts continued into later writings: the third edition of *Against Method* repeats the claim that '[t]he arts ... are not a domain separated from abstract thought, but complementary to it and needed to fully realise its potential' (Feyerabend 1975/1993, p. 267). But into those later writings, one also finds a new humanistic defence: the arts offer those creative and imaginative resources and outlets vital to at least certain forms of human flourishing – a claim already teased in the closing sentences of 'Science, Freedom, and the Good Life', which proposes 'the preservation of human happiness' as a 'unifying ideal', which Feyerabend hoped could repair the 'fragmentariness' of the arts and sciences (Feyerabend 1968, p. 134). The humanistic defence of the arts seems to be motivated by an existential concern – to offer people 'a survey of the possibilities of human existence' (Feyerabend 1991, 495). The whole range of human epistemic, artistic and cultural resources ought to be brought to bear to help human beings to fully and happily 'participate in the richness of being and [achieve] a more tolerant and compassionate view of how the sciences, arts, and religions could jointly contribute to improving the human condition' (Oberheim 2006, p. 598).

The final aspect of the critique of scientific modernity concerns *disenchantment*, in the sense made famous by Max Weber. At its most general, the concern is that a scientific picture of the world, once internalised and institutionalised within a society, generates or exacerbates a sense of profound alienation. Whether in itself or due to its contrast with earlier religious visions of the world, what the modern scientific *Weltbild* offers is a certain conception of our natural situation within the wider order of nature that is existentially impoverished, unable to nourish our inveterately teleological needs to find meaning and purpose within our activities. Strictly speaking, the real problem is not a scientific vision of the world itself, but rather a certain form of scientistic naturalism – the doctrine that what ultimately, fundamentally exists is the world as described by the modern sciences, within which values, meanings and purposes are subjective projections of our consciousness, rather than genuine features of reality that might act as measures for the meaningfulness of our activities.

Robert Farrell, one of the few Feyerabend scholars to explore this theme, argues that the deep concern is with reductionistic visions of the world, that 'occlude' the abundance of the world by denying or downgrading certain aspects of lived experience integral to *human* life:

> Feyerabend is highly critical of unified world-views *when they are reductionistic in character:* when they achieve unity at the expense of denigrating large sections of reality as not really real; where mind, or culture, or aesthetic experience, or whatever aspects of existence which resist reduction are perceived as illusory and metaphysically second-rate. (Farrell 2003, p. 234)

Such concerns become prominent in the very late writings, such as *Conquest of Abundance*, with its guiding worry that adoption of a scientistic picture of the world erodes the possibility of a 'meaningful existence', by situating us within a 'cold, austere' worldview, apt to leave us 'scattered, aimless' (Feyerabend 1987, pp. 13, 5, 246).

Concerns about homogenising violence, philistine scientism and the existential disenchantment of human life are evocative and dramatic, especially when brought together in the deeper historical vision of human condition sketched in *Conquest of Abundance* and *Naturphilosophie*. But they are also a mixed bag.

Before criticising them, three conciliatory comments are in order. First, the critique of scientific modernity is consistent with the critique of science's self-understanding: inflated estimations of the epistemic capacities of the sciences might, for instance, fuel derogatory attitudes towards other forms of knowledge and understanding. Second, the critique of scientific modernity resonated with the intellectual and cultural context in which Feyerabend's writing was consistent with what Oberheim calls his 'remarkable ability to adapt to changing interests and attitudes' (Oberheim 2006, p. 24). These included, *inter alia*, the emergence of the environmentalist movement, the post-colonial critiques fired by decolonisation, and the anti-technological counterculture heralded by E. F. Schumacher's little essay, *Small is Beautiful*. Underlying this was, perhaps, what John Preston insightfully calls Feyerabend's 'reactionary romanticism' – a nostalgic yearning for a simpler way of life, unmarked by the abstractions of intellectuals, the rapacity of our hyperactive consumerist world and the existentially depleted scientific vision of the world (Preston 2000, p. 621).

A third feature of the critique of scientific modernity are its continuities with themes and moods in the early-twentieth century Austrian intellectual culture that were home to Feyerabend. Where earlier scholarship painted him as an idiosyncratic analytical philosopher, newer studies of his work explore his status as *ein Philosoph aus Wien*, to cite the title of an important collection edited by Friedrich Stadler and Kurt R. Fischer (2006). Perhaps the best example of Feyerabend's debts to early Austrian philosophy is the influence of Wittgenstein, whose own criticisms of scientism resonate through the critique of scientific modernity (Kidd 2017a). Think of the warnings scattered throughout *Culture and Value*, that the hegemony of science is 'a way of sending [us] to sleep', that modern people 'think that scientists exist to instruct them' and artists only to 'entertain', and that 'the age of science and technology' may prove to be 'the beginning of the end for humanity' (Wittgenstein 1980, pp. 5, 36, 56).

Considered in wider context – biographical, intellectual and cultural – one can easily understand why Feyerabend developed a critique of scientific modernity. Moreover, a sensible claim lies at its heart: the beneficent potential of science is contingent, not guaranteed, since it can only serve our practical and epistemic purposes when properly understood and carefully managed. But that is difficult, requiring a constant critical scrutiny, informed by a clear set of purposes that the sciences ought to serve, and which ought to emerge from robust critical reflection about the 'good life' – a point where Feyerabend really ought to have engaged with pragmatism, for some of the reasons sketched by Brown (2016). The chatty essay 'How to Defend Society against Science' serves as a good statement of Feyerabend's considered position that there is 'noting inherent in science or in any other ideology that makes it essentially liberating', since any ideology can 'deteriorate', becoming 'rigid' and 'oppressive' (Feyerabend 1975c, p. 3).

Such conciliatory remarks on context and motivation can help us to better appreciate why so much of Feyerabend's later work was devoted to a critique of scientific modernity. The big question is the sophistication and success of that critique – or, perhaps, the three critical themes of homogeneity, philistinism and disenchantment. My judgement is that the critique of scientific modernity was far less impressive than its sister critique, for reasons that it was unsystematically developed, poorly integrated with the other critique, and underinformed by existing resources and ideas.

In my judgement, three main issues contributed to the deficiencies of Feyerabend's critique of scientific modernity. The first is a failure, on his part, to engage with sub-disciplines and traditions whose concerns and insights pertain to topics integral to critique of scientific culture. Consider, for instance, the fact that the concerns about epistemic and cultural homogenisation would be far more effectively theorised using theoretical frameworks sensitive to the relationships between epistemic diversity and social power. Yet there is no engagement at all with feminist and social epistemology (at the level of projects of enquiry) or post-colonial theory (at the cultural and political level). Consider, too, the salutary desire to better organise scientific activities to support our practical, epistemic and social interests. It is well-taken, but relies on slogans about 'human happiness and liberty' – usually harping on themes from John Stuart Mill (see Lloyd 1996) – rather than proper engagement with pragmatism and political philosophies. By ignoring such potential allies, several substantive weaknesses were built into the critique of scientific modernity, most obviously in the areas of social epistemology and normative political philosophy.

Second, the failure to acquire concepts and methods necessary for the articulation of certain styles of critique, which is clearest in the 'disen-chantment' theme. Feyerabend's remarks rely on a generic vocabulary of 'meaninglessness', 'aimlessness', and so on, coupled to ominous warnings of the dire existential consequences of the 'conquest of abundance'. But, no effort is made to elevate such rhetoric by providing it with substantive philosophical support, for which the most obvious source is existentialism and phenomenology. The later writings of Martin Heidegger offer acute diagnosis of the 'distress' of this, our 'destitute age', suffering through the 'oblivion of Being', owning to the entrenchment of 'technological' ways of revealing the world that have, almost entirely, 'driven out' all other possible 'ways' (Heidegger 1971, p. 91; 1977, p. 27f.). Such luminaries of existential phenomenology offer acute criticisms of the existential and cultural con-sequences of sophisticated forms of scientism, although they barely figure in Feyerabend's writings. But, this is seriously problematic since, without their conceptual resources, talk of the 'disenchantment' of the world remains intolerably bland.

The third source of the weaknesses of Feyerabend's critique of scientific modernity is closely related: the failure to situate this aspect of his work within the Continental European tradition. As Peter Strawson once remarked, 'more or less systematic reflection on the human situation', of a sort found in the Continental European traditions and aimed at affording 'a new perspective on human life and experience', belongs to a 'species of philosophy' that is 'quite different' from the analytical one (Strawson 1992, p. 2). I doubt that Feyerabend can really be classified as an analytical philosopher, as by its criteria of precision and argumentative rigour, he often comes out badly. But there is a good case to be made for seeing him as being closer to the Continental European 'species' – think of his evident concerns with reflection on the conditions for meaningful human life, coupled to an ardent critique of contemporary forms of life, animated by acute suspicion about the historically contingent epistemic authorities of late modernity, most obviously the sciences.

Looked at this way, Feyerabend, at least when engaged in the critique of scientific modernity, seems much closer to critical theory and the Frankfurt School and to the later writings of Heidegger and Husserl, than to the dominant themes and figures of analytical philosophy. Trying to make sense of the 'aimlessness' and 'destruction' wrought by the 'conquest of abundance' is easier, if one appeals to *Dialectic of Enlightenment, The Question Concerning Technology* or *Crisis of the European Sciences*. Such works openly explore such themes as the complicity of scientific modernity in the 'world-domination of nature', existential 'distress' and a philistine 'barbarian hatred of spirit'.

I have only sketched out some obvious points of contact between Feyerabend's later writings and such complex movements and traditions as social epistemology, feminist and post-colonial philosophies, pragmatism, critical theory and existential phenomenology. The current scholarship has been slow to explore and assess those connections, albeit with a few recent honourable exceptions, doubtless a reflection of the demographics of Feyerabend scholars, most of whom work in the history of analytic philosophy of science. But, the outstanding question is *why* Feyerabend himself did not avail himself of the rich resources of those traditions, all of which were well developed during the later years when he was experimenting with the critique of scientific modernity. In a few cases, it seems there was acquaintance with little engagement; for instance, Feyerabend read Husserl's *Crisis of the European Sciences*, which he praised as a 'remarkable essay' (Feyerabend 1987, p. 274), but frustratingly confines himself to the blunt comment, 'Trouble started in antiquity' (Feyerabend 1999, p. 253). But there is no good evidence of any substantive engagement with feminist, post-colonial, pragmatist or existentialist philosophies or with phenomenology or critical theory – only a few references, here and there, to Nietzsche, Kierkegaard,

Derrida and a few others, such that one has to go digging to discern the extent and nature of their influence upon him (see, e.g. Kidd 2011).

My judgement is that Feyerabend's ventures into a critique of scientific modernity could not be developed adequately with the methods and conceptual resources available from Anglophone philosophy of science. Since cultural critique was not a theme of that 'species' of philosophy, that is not in itself a problem, since those resources were available from other traditions, of which we must believe Feyerabend was aware. His failure to engage with them is therefore problematic, an important consequence of which is that we must rethink his vaunted intellectual eclecticism. Oberheim rightly praises Feyerabend as a philosophical pluralist, an opportunistic thinker, actively concerned to seek and exploit new ideas and ways of thinking 'to improve our understanding through immanent criticism of existing points of view, to counter conceptual conservatism, and thereby to help promote the critical development of new points of view' (Oberheim 2006, p. 287). I agree, but with the caveat that, in practice, what we find is really a sort of *patchwork pluralism*. Feyerabend's main loves are, in practice, classical Greek scholarship, the history of physics and Renaissance art. Such eclecticism is impressive in its scope and imaginative in its use, although critics have queried the rigour of the scholarship of ancient Greek history, literature and art (see Clark 2002; Preston 2016). But such eclecticism falls short of what is really needed for a critique of scientific modernity – namely, substantive engagement with the feminist, post-colonial and other sources so patently absent from the later writings.

9.5 Conclusion

The aim of this chapter is that Feyerabend's later work is best understood as a series of experiments with an ambitious project of philosophical anti-scientism – a 'critique of scientific reason'. As a venture into normative philosophy of science, it offers an attractive vision of the ways that philosophy can play an important social role in ways taken up by later developments in the discipline. An obvious case is what Carla Fehr and Kathryn Plaisance call 'socially relevant philosophy of science', focussed on collaborative relationships with scientists, addressing questions of policy and regulation, investigating the relationships among scientific and non-scientific communities and assessing the efficacy of the norms, practices and structures of philosophy of science (Fehr and Plaisance 2010, p. 314). But appealing for certain kinds of philosophical work is not the same as actually doing it – a fact that might prompt the question of how, at the end of the day, we should understand the contemporary significance of Feyerabend's work.

A downbeat answer is that the 'critique of scientific reason' was a prescient appeal for a broadened conception of the aims and methods of philosophy of science, albeit one subsequently put into practice by feminist philosophers of science, social epistemologists, and post-colonial science and technology studies scholars. By breaking the strictures of positivist visions of science, Feyerabend was instrumental in opening a space that was later more thoroughly occupied and developed by others, which has the effect of assigning him a greater place in the *history* of philosophy of science, but not its present and future course. To use his own backhanded compliment to his discipline, Feyerabend would emerge as a philosopher with a 'great past'.

This downbeat judgement on Feyerabend's work seems, however, too modest, on two counts. First, there is more going on in his writings than simply a critique of positivism, as evidenced by the topical scope of the critique of scientific reason – think of its inclusion of educational, cultural and existential concerns. Assessing Feyerabend's significance through the agenda of contemporary philosophy of science leaves us apt to miss much of the variety and richness of his kaleidoscopic interests and concerns. Second, there are many original, important philosophical ideas in the later writings, many currently still underexplored. Think of Martin Kusch's discussion of the distinctive forms of ontological and political relativism within the later writings (Kusch 2016); the application by Helene Sorgner of Feyerabend's ideas on the democratic control of science to criticise contemporary claims about scientific expertise (Sorgner 2016); the sturdy literature devoted to the later Feyerabend's accounts of scientific realism and perspectivism, represented by Matt Brown (2016), Ronald Giere (2016) and Luca Tambolo (2014). Moreover, careful study of the later writings can offer other things, such as the components for a robust doctrine of the historical contingency of the sciences (see Kidd 2016d, 2017b).

Going beyond hackneyed images of Feyerabend as 'the worst enemy of science', an appreciation of his critique of scientific reason shows him to be an engaged philosopher of science with ameliorative aspirations. Feyerabend emphasises the 'messiness' of science, its complexity, contingency and entanglement with other areas of human life. He also calls to our attention the many dangers arising when this messiness is either forgotten or ignored due to our dogmatism, complacency, critical lapses, or our falling victim to the epistemic traps set by scientistic 'myths'. Without trying to guess Feyerabend's views, my own sense is optimistic. Many contemporary movements in philosophy of science have been in the direction of this sort of project. If that is so, perhaps much of philosophy of science today is, to the surprise of many, strikingly Feyerabendian.[1]

[1] Acknowledgements: I am very grateful to the editors Karim Bschir and Jamie Shaw for their invitation and comments, to an audience at the Science and Technology Studies Department at University College London for discussion, and for the ongoing enthusiasm and ideas of Chiara Ambrosio, Matt Brown, Robin Hendry, and John Preston.

CHAPTER 10

Against Expertise
A Lesson from Feyerabend's Science in a Free Society?

Matthew J. Brown

10.1 Introduction

As my title suggests, I will examine a lesson from Feyerabend's controversial work, *Science in a Free Society*: that free societies should not invest scientific experts with special epistemic or social authority. At least, I will ask, should we take this claim from Feyerabend as a lesson? In particular, I see Feyerabend's argument about expert authority as a substantive challenge to a central commitment of many philosophers of science who reject the ideal of value-free science, a commitment to *the ineliminability of expert judgment*. An argument against this principle is articulated clearly and forcefully, if somewhat roughly, in both his *Science in a Free Society* (1978, hereafter *SFS*) and a related article, "How to Defend Society against Science," originally published in *Radical Philosophy* (1975c, hereafter *HDSS*).

Science in a Free Society, published over forty years ago, was Feyerabend's least well-received work, one even Feyerabend himself came to dislike. In his later years, he apparently wished that the publisher would cease reprinting the book.[1] There are a number of reasons this might have been. There is the problematic, even naive political philosophy of "the free society" contained in the book. The book defends a form of "relativism," a term Feyerabend tried to dissociate himself from in his later years. The third part of the book collects a rather testy set of responses to critics, which Feyerabend may have come to see as too salty even for his taste. And yet, major parts of the book's arguments were incorporated in the second and third editions of *Against Method*, Feyerabend's most well-known work, one that he continued to carefully rework throughout the middle and later periods of his career. The core ideas of *SFS* remained important to Feyerabend.

First, I provide some background for this discussion by looking at contemporary discussions of values in science and the role of expert

---

[1] The book remains in print to this day.

judgment in those discussions. Here, I lay out the importance and rationale behind the commitment to the ineliminability of expert judgment. Then, I briefly discuss Feyerabend's problematic political philosophy, in order to disentangle its commitments from what is potentially of value in his argument about the role of experts in society. Next, I lay out four theses about the relation between citizens and scientific experts that forms a major part of Feyerabend's arguments in *SFS* and *HDSS*. After briefly discussing a major caveat that Feyerabend makes about these claims in *SFS*, I evaluate Feyerabend's argument and its bearing on the question of the eliminability of expert judgment, ultimately arguing that Feyerabend presents a dilemma for the role of experts in society that we cannot avoid, though hopefully we can find a compromise position between its horns.

10.2 Background: Expert Judgment and Values in Science

Before we start looking at Feyerabend, I want to review some common commitments in the science and value literature which give contemporary relevance to Feyerabend's arguments. Although these commitments are common in the literature, for the sake of space, I will largely focus on arguments about *inductive risk* presented by Heather Douglas (2000, 2009). Douglas begins from the fact that ampliative claims are pervasive in scientific inquiry – claims that go beyond what is strictly implied by the evidential basis for those claims. Douglas argues that all such ampliative claims have a risk of error, and that the consequences of error can include social and ethical consequences. Because scientists have the ordinary responsibilities to consider the consequences of their actions, whenever scientists make empirical claims or inferences, they ought to make social or ethical value judgments to weigh the consequences of potential errors. You can run the same kind of argument with research questions, choice of language, or concepts with which to frame data or hypotheses, theory choice in the broad sense, or various other consequential contingencies in science (see Brown, 2020, Chapter 2). The conclusion of each argument is that scientists ought to exercise (non-epistemic) value judgment.

A common response to arguments like Douglas's is the *deferred decision response* – scientists need not make the risky empirical claims, but only pass on the relevant information to make those claims to decision-makers, who rightly make the relevant value judgments.[2] Or, alternatively, that

---

[2] Various versions of the response can be found in Jeffrey (1956); Mitchell (2004); Pielke Jr. (2007); Betz (2013); Edenhofer and Kowarsch (2015). The concept of the "deferred decision response" was

scientists accept hypotheses for the purposes of scientific *belief* only, in such a way that is completely cut off from *action*. Scientists' proper work can thus remain value-free, and the burden of judgment is passed on to the relevant decision-makers. For example, Sandra Mitchell explicitly gives the deferred decision response to Douglas. First, she describes Carl Hempel's own approach to inductive risk. Hempel distinguishes between the inductive risk of hypothesis acceptance in science, which depends on purely epistemic values, and inductive risk in "practical contexts ... when the hypothesis is to form the basis of action or policy," where social or ethical values are appropriate. On this basis, she responds to Douglas thus:

> Douglas justifies her 'expansion' of Hempel's argument from inductive risk into the domain of deciding which theory to accept as true by appealing to the authority of science in our society ... This conflation of the domains of belief and action confuses rather than clarifies the appropriate role of values in scientific practice. Indeed, to make public one's belief that a given hypothesis is true is an action, and in certain contexts a scientist might judge that stating what he or she is scientifically warranted to believe is politically inadvisable ... The values appropriate to generating *the belief* and the values appropriate to generating *the action* are different. (Mitchell 2004, pp. 250–251)

Mitchell goes on to argue that we must analytically separate the two functions, even when the same particular individuals might be involved in doing the science and making the policy decisions.

There are a number of problems with the deferred decision response. The deferred decision response usually focuses only on the final decision to accept or reject a hypothesis.[3] But there is a regress of value-laden decisions throughout scientific inquiry, such as decisions about framing a hypothesis to test and about methodological choices. Also "the relevant information" that scientists might provide policymakers itself consists of further risky ampliative empirical claims, such as the characterization of evidence or attributing a probability to various hypotheses.[4] The rigidly defined roles

---

first worked out collaboratively with my co-author Joyce Havstad, and she deserves the lion's share of the credit for it. See Havstad and Brown (2017).

[3] The most nuanced version of the deferred-decision response appears in the "pragmatic-enlightened model" for science-based policy by Edenhofer and Kowarsch (2015). This version traces the role of value judgment through all of the decisions made throughout the process of scientific inquiry, setting up a "cartography" of pathways through the science that policymakers can decide how to navigate based on their value judgments. For critique of this version of the response, see Havstad and Brown (2017).

[4] Richard Rudner already demonstrated this problem for the deferred decision response in Rudner (1953). Dan Steel refers to the uncertainty about probability estimates as "second-order uncertainty" (2016).

and oversight that the deferred decision response would require would likely *stifle* scientific progress. It is plausible that only the scientific experts have the *competency* – the knowledge, skills, and experience – necessary to adequately anticipate and weigh consequences of error. Finally, in practice, the role of scientific research and science advising are often *indistinguishable*.

These objections to the deferred decision response amount to a view that is a near consensus among those who reject the value-free ideal: the ineliminability of expert judgment. Simply put, experts must make value-laden scientific decisions with an eye to guiding policy- and decision-makers. I have defended this claim myself (Havstad and Brown 2017; Brown 2020), and it is important to the ways that those of us who have been defending the value-ladenness of science have been working. However, this position also raises problems – of accountability, trust, and legitimacy of experts in a democratic society (as many defenders of the value-free ideal point out, and as many advocates for values in science attempt to address; see Betz 2013; Bright 2018; Douglas 2005, 2009, Chapters 7–8). The ineliminability of expert judgment is the thesis that I want to use Feyerabend's argument to question.

10.3 Feyerabend's Problematic Political Philosophy

Before considering Feyerabend's arguments that, I think, call into question the ineliminability of expert judgment, it must be recognized from the outset that Feyerabend's theory of the "free society" in *SFS* is notoriously problematic. Among other things, it is committed to a kind of naive and simplistic libertarianism, problematic views about group identity and the nature of cultural traditions, and of course relativism. The untenability of Feyerabend's ideal of a "free society" has been so thoroughly discussed elsewhere that I do not need to recapitulate the problems (see, e.g., Munévar 1991; Brentano 1991; Kidd 2016a). However, much of *SFS* and *HDSS* are concerned not with defending this problematic approach, but with determining the role of science in this free society. Much of Feyerabend's account of science in society can be retained without adopting the problematic picture of a free society. In fact, few premises in the argument about science in society depend on the problematic aspects of Feyerabend's political philosophy.

10.4 Feyerabend on the Role of Science in a Free Society

Let us see, then, if we can understand Feyerabend's argument as one about the role of science and scientific experts in the kind of "free society" that is much more widely endorsed, one committed to basic liberal–democratic values and to political processes that both respect the rights of the citizens while being likely to deliver good outcomes on matters of public interest.

Even on this background, Feyerabend can make a case for an approach to scientific expertise that denies a special epistemic and cultural authority to the scientist. Feyerabend articulates and defends four increasingly radical claims about science and its place in society that, together, point toward a decentering of experts in society and a rejection of expert authority.

10.4.1 Citizens Can and Should Evaluate Expert Opinions

Feyerabend holds that citizens have the right to evaluate expert opinions for themselves, and that they ought to exercise that right. Here he does not simply mean that citizens have a right to believe what they will, in a purely private sense of their personal opinions. He means that every citizen has a right to evaluate, criticize, and reject in whole or in part expert claims for the purposes of public decision-making. He has three main arguments for this view.

First, citizens should be free to evaluate expert opinions for themselves for broadly Millian reasons, that freedom and plurality of opinion and open debate leads to better epistemic outcomes. Feyerabend adopts wholesale the arguments of John Stuart Mill's *On Liberty* and takes them, perhaps, further than Mill himself.[5] Feyerabend sees a plurality of competing views in society, with the freedom to advocate for any belief and to adjudicate competing beliefs for oneself, as tending to increase the justification of those beliefs, the likelihood that true beliefs are found and adopted, and the very meaningfulness of those beliefs (SFS, p. 86). This putatively Millian argument is familiar to Feyerabend readers and much commented upon (Lloyd 1997; Staley 1999). Feyerabend uses the argument here to defend the process of citizens evaluating putative expert opinions for themselves, as being more likely to lead at least some citizens to adopt true beliefs, as well as their actually being justified in so adopting them and knowing what the beliefs actually mean. Feyerabend argues, with Mill, that this freedom of belief and expression is more likely than deference to others to lead to human progress.

[5] See Jacobs (2003) for a critique of Feyerabend's interpretation of Mill.

Second, Feyerabend argues that *even if* citizens having the last say over the claims of experts leads to worse *outcomes* for themselves or for society, it is nevertheless justified, as it would contribute to human freedom and to the development of a mature democratic citizenry – these goods outweighing the value of mere true belief or the risks of policymaking without the benefits of expert assessments. On the one hand, Feyerabend takes this as just definitional of democracy – in a democracy, the public must participate in fundamental decisions; they cannot be left to the experts. He simply asks us to be fully consistent on this point. Feyerabend also argues that the learning and matura-tion necessary for citizens to become wise or skillful public participants requires being that they be allowed to try and, perhaps, do poorly. According to Feyerabend, "[p]articipation of laymen in fundamental deci-sions is therefore required *even if it should lower the success rate of the decisions*" (SFS, p. 87).[6] It is the kind of thing that distinguishes democracy from totalitarianism, which is free to focus on whatever is most "effective."

However, Feyerabend also thinks that motivated "laymen," in fact, can be competent enough to make good decisions regarding scientific information: "science is not beyond the natural shrewdness of the human race" (SFS, p. 98). Scientists, of course, are mere human beings; no miraculous feat transforms a human into a scientist. Among Feyerabend's favorite evidence for this point is the way that a skilled litigator can digest and critically analyze expert testimony on the fly, and thus expose weaknesses and uncertainties. Without specialized training, (suitably dedicated) non-experts can evaluate scientific information competently for themselves. The ability of historians and philosophers of science to analyze scientific publications and archival records, to break down the decisions behind them, and to critically assess them, is further evidence in Feyerabend's favor.

10.4.2   Citizens Can and Should Supervise Science

Feyerabend goes one step further: not only should citizens decide for themselves whether to believe and how to use scientific information; they should regularly subject science to careful scrutiny. He even goes so far as to suggest that citizens should "supervise" science.

---

[6] There is reason to object to the use of the term "laymen" or "laity" as distinct from experts precisely because it sets experts up as clergy-like authorities. But science is not a church, and scientists are not authorities ordained by a higher power. I suspect Feyerabend uses this term knowingly and tongue-in -cheek. I will typically use "citizen" or "non-expert" instead of "laymen," except when paraphrasing Feyerabend's own use of the latter.

> Laymen can and must supervise Science ... it would not only be foolish but downright irresponsible to accept the judgment of scientists and physicians without further examination. If the matter is important, either to a small group or to society as a whole, then this judgment must be subjected to the most painstaking scrutiny. (SFS, p. 96)

According to Feyerabend, we should elect committees of non-experts to regularly subject scientists and their work to review before it is put to social use. He specifically mentions reviewing evidence for theories before they are taught, reviewing the safety of nuclear power plants, and reviewing the efficacy of scientific medicine against alternatives.

One of the major reasons that Feyerabend sees the need for citizen supervision is his view that scientists themselves are often prejudiced and untrustworthy:

> Expert Opinion [is] often Prejudiced, Untrustworthy, and in Need of Outside Control ... scientists quite often just don't know what they are talking about. They have strong opinions, they know some standard arguments for these opinions, they may even know some results outside the particular field in which they are doing research but most of the time they depend, and have to depend (because of specialization), on gossip and rumours. (SFS, pp. 88–89)

The argument here is a bit complicated. One might balk at the claim that scientists "don't know what they are talking about." But in this context, it is clear that Feyerabend is worried about the rather common phenomenon of scientific experts speaking authoritatively outside of their (narrow) area of expertise. Feyerabend spends a lengthy section of *SFS* criticizing "Objections to Astrology: A Statement by 186 Leading Scientists," a brief statement published in *The Humanist* magazine, September/October 1975 issue (see also Kidd 2016b). He does this not because he cares to defend astrology, but to show the irresponsible attempt of this band of scientists to assert their authority. He shows that many of the scientists knew nothing about astrology and did not have expertise directly relevant to the issue. Feyerabend summarizes, "[t]hey neither know the subject they attack, astrology, nor those parts of their own science that undermine their attack" (SFS, p. 92). He demonstrates that by the lights of mainstream science, some of the claims from the statement were baldly false, while others are irrelevant.[7] These are in themselves good reasons to insist on some increased scrutiny, to check on

---

[7] Feyerabend was in good company, here. Astronomer, science popularizer, and pseudoscience critic Carl Sagan similarly objected to the content of the statement in a letter to the editor of *The Humanist*.

whether experts are advising within or outside of their area of expertise. But should they also be supervised within their area of expertise?

One reason to think so comes from a common view about the nature of (scientific) knowledge in philosophies of science like Feyerabend's, Kuhn's, and Lakatos's. On this view, science contains some necessarily presumptive, dogmatic element that both makes scientific progress possible but also contingent (different presumptions would have led to progress in a different direction). As Feyerabend puts it,

> Such ideas are not simply errors. They are necessary for research: progress in one direction cannot be achieved without blocking progress in another. But research in that 'other' direction may reveal that the 'progress' achieved so far is but a chimera. It may seriously undermine the authority of the field as a whole. Thus science needs both the *narrowmindedness* that puts obstacles in the path of an unchained curiosity and the *ignorance* that either disregards the obstacles, or is incapable of perceiving them. (SFS, p. 89)

That functional "ignorance" can be provided by non-expert control of science. But this way of putting things (" … science needs … ") is inapt. For Feyerabend's argument in *Against Method*, such a claim is on point: dilettantes, amateurs, and heterodox scientists have much to offer *science* because they can pursue paths closed off by scientific orthodoxy, and thus promote the pluralistic growth of knowledge. But the appropriate question in the line of argument Feyerabend is pursuing in Part 2 of *SFS* is not, "What does science need?" but "What does society need?" The nascent point that Feyerabend is making is rather that, precisely because scientific orthodoxy may block the development of knowledge in precisely the direction that some citizens may desire, while producing progress in directions that might turn out to be chimerical, we should not just take the scientists' word for it. We must supervise their work to determine whether this is the case.

Even more narrowly construed, within their proper area of expertise, and according to the paradigm or research tradition adopted, Feyerabend raises concerns about the trustworthiness of scientists, and insists that they must be supervised by citizens. Even in this area, Feyerabend thinks that scientists can hide "uncertainty, indefiniteness, the monumental ignorance" behind jargon and assertions of epistemic authority (SFS, p. 98). It may not be too much of an exaggeration to say that scientists regularly oversell the strength of their knowledge to society, in ways that hurt their credibility. For instance, Jim Brown and Jacob Stegenga have argued forcefully, the trustworthiness of contemporary medical science is in grave doubt (J. R. Brown 2002, 2016;

J. R. Brown 2008a, 2008b; Stegenga 2018). The human and social sciences are beset by replication crises (Smaldino and McElreath 2016). Arguments have been made from within science that most published findings are in fact false (Ioannidis 2005). And even where the technical quality of the research is good, if science is value-laden, and scientists incorporate their personal values into science, this may give reason for certain groups to distrust those results, if they disagree with said values. Without any established democratic means for influencing scientists' value judgments, this issue is pressing.

10.4.3 Science Is Just Another Ideology or Interest Group

Feyerabend argues that we should see scientists as purveyors of just another ideology, that is, as a group with a specific perspective and a characteristic set of beliefs reflecting that perspective (see Selinger 2003, 360–361). This can be seen, for instance, in the dogmatic form of science education:

> Scientific 'facts' are taught at a very early age and in the very same manner in which religious 'facts' were taught only a century ago. There is no attempt to waken the critical abilities of the pupil so that he may be able to see things in perspective. At the universities the situation is even worse, for indoctrination is here carried out in a much more systematic manner. (HDSS, p. 4)

This can be contrasted with Kuhn's view of science education. While Kuhn agrees that science education is rigid and dogmatic, he argues that this is necessary for providing scientists with the working commitments that make progress possible (see Kuhn 1963). Feyerabend sees this as evidence rather that scientists favor their own lore and wish to use their special place in society to push it as universal truth on unsuspecting young minds.

Feyerabend taught at the University of California, Berkeley during the desegregation of public education in the United States and the attendant increase in non-white enrollments at Berkeley, an experience that had a profound effect on his philosophical outlook (Kidd 2013, pp. 408ff; 2016a, p. 124). He came to see his educational role as essentially oppressive, pushing "reflections of the conceit of a small group who had succeeded in enslaving everyone else with their ideas," and to find the very idea revolting (SFS, pp. 118–119). The problem was that the ideology of the privileged remained centered:

> But equality, racial equality included, then did not mean equality of traditions; it meant equality of access to one particular tradition – the tradition of the White Man. (SFS, p. 76)

On Feyerabend's view, "the tradition of the White Man" was centered because of the power of white supremacy, and not because of the undeniable superiority of their results or method:

> ... the comparative excellence of science has been anything but established ... Science does not excel because of its method for there is no method; and it does not excel because of its results: we know what science does, we have not the faintest idea whether other traditions could not do much better. So, we must find out. (SFS p. 106; cf. HDSS, p. 5)

Because science is just another ideology, and scientists are just an interest group promoting said ideology, historically through the exercise of colonial, patriarchal, and white supremacist power, there is no justification in giving science greater authority than any other tradition. Hence, citizens will have to make up their own mind about whether to accept scientific conclusions, and where science might have significant impact on society, it must be supervised by citizens. The value-ladenness of science strengthens this point, as there exist no established means for scientists to incorporate any but their own personal values into their scientific decisions.

Feyerabend's position here is supported by a philosopher of a very different temperament, the great American philosopher of science, education, and democracy, John Dewey:

> A class of experts is inevitably so removed from common interests as to become a class with private interests and private knowledge ... No government by experts in which the masses do not have the chance to inform the experts as to their needs can be anything but an oligarchy managed in the interests of the few .... (Dewey 1927, pp. 364–365)

Dewey rejected the technocratic role of scientific experts in governance, all the while insisting on the importance of science. For Dewey, however, the participation of the public in the process was paramount. Any opportunity to exercise authority, uninformed and uncontrolled by the public, amounts to autocratic rule by private interests.

10.4.4   Science Should Be Separated from the State

Another consequence which Feyerabend takes from the claim that science is a culturally specific ideology or a particular interest group is that it should be treated with a certain distance by the state. Indeed, Feyerabend claims that there should be formal separation between science and the state, modeled on the formal separation between church and state that is characteristic of many liberal democracies:

The most important consequence is that there must be a formal separation between state and science just as there is now a formal separation between state and church. Science may influence society but only to the extent to which any political or other pressure group is permitted to influence society. Scientists may be consulted on important projects but the final judgement must be left to the democratically elected consulting bodies. (HDSS, p. 6)

This may seem to conflict with Feyerabend's second claim, that society should supervise or monitor science. This points to two ways of interpreting the claim that citizens should supervise science. One meaning of "supervise" is control; a manager who supervises an employee's work is controlling the work of that employee. This generates a tension for Feyerabend: if science is a private activity, separate from the state, then non-scientist citizen control of that activity seems to violate a host of liberal-democratic rights such as freedom of association and freedom of opinion. Given Feyerabend's insistence on the rights of individuals and groups throughout *SFS*, this seems at odds with the overall argument.

Another meaning of "supervise" is to monitor and evaluate without interfering. This seems more consistent with Feyerabend's project. Here, citizens are charged with closely scrutinizing what science does, insofar as they may want to let science influence society on particular points. This is compatible with formal separation of science and state. In a liberal democracy with strong separation of church and state, religious citizens can vote, religious leaders are consulted on public decisions, and sometimes even religious organizations can be contracted to fulfill state functions, such as providing homeless shelters. But in the latter two cases, it is up to citizens or democratically elected representatives to independently supervise by evaluating the claims of religious leaders and the functioning of religious organizations. Similarly, if science is to be thought of as a private activity, but allowed to influence society, it must be monitored and independently evaluated.

10.4.5 *Summary of the Argument*

Feyerabend has argued as follows. Scientists often speak outside of the areas of their expertise with the same confidence and assertion of authority as within their area of expertise. They do this even when they have not properly looked into the question at hand. Even within their area of expertise, scientists tend to be close-minded and arrogant. The very nature of scientific knowledge requires a narrow-mindedness that makes it a problematic resource for society at large. In effect, science is an ideology or an interest group, with its own articles of faith, its own specific aims and values that may be at odds with those of citizens. What's more, we cannot overlook these issues on the basis of the superior results that science has delivered, because there has never been a fair and even playing field from which to adjudicate such superiority. Science should thus not be granted any special epistemic authority, and citizens will have to make up their minds for themselves whether to believe the deliverances of science. Science should be formally separated from the state, not given a special role in policy or education. That does not mean science cannot or will not be consulted; only that it does in the way any interest group operates in a democratic society. Whenever science may influence or impact society, it should be monitored and evaluated by non-experts, to ensure its influence on society is beneficial and legitimate.

10.4.6   The Maturity Caveat

Feyerabend occasionally qualifies the sorts of claims made above as applying to a society of "mature people," and he quotes John Stuart Mill who argues that his views on pluralism and free exchange of ideas "is meant to apply only to human beings in the maturity of their faculties" (SFS, 29n). Near the end of his main argument, he makes the following major caveat:

> The separation of state and science (rationalism) which is an essential part of this general separation of state and traditions cannot be introduced by a single political act and it should not be introduced in this way: many people have not yet reached the maturity necessary for living in a free society ... The maturity I am speaking about is not an intellectual virtue, it is a sensitivity that can only be acquired by frequent contacts with different points of view. It can't be taught in schools ... But it can be acquired by participating in citizens initiatives. This is why the *slow* process, the *slow* erosion of the authority of science and of other pushy institutions that is produced by these initiatives is to be preferred to more radical measures: citizen initiatives are the best and only school for free citizens we now have. (SFS, p. 107)

This can easily be seen as taking the sails out of the radical claims canvassed earlier. We thought Feyerabend was recommending that citizens make up their own minds and that science be formally separated from the state, but he is really recommending neither. Should science thus retain its epistemic authority and special place with respect to the state?

Feyerabend is not entirely clear here, but there are reasons not to read this caveat as complete reversal but rather as a plea for gradualism and bottom-up, rather than top-down change. He emphasizes here the importance of informal education and the encounter with other points of view, some of which will be "non-Western" or non-scientific perspectives. Also, in a passage partially quoted earlier, Feyerabend tells us that the relevant kind of maturity is not learned in schools, but "by *active participation* in decisions that are still to be made" (SFS, p. 87). Cultivating maturity means allowing citizens to decide for themselves, as quoted above, "even if it should lower the success rate of the decisions." However, current citizens should not all at once be thrust in to having decide everything for themselves. They should gradually do so through growth of participation in "citizens initiatives."

We live in a world where many gullible citizens are liable to be hornswoggled by industry-driven merchants of doubt into believing that climate science is a hoax; by QAnon conspiracists into believing that Donald Trump is working to secretly save the world from the secret Satanic pedophiles who secretly run it, and by nutrition gurus who hold the secret to the miracle of health but choose to release it as a multi-level marketing scheme rather than through scientific or even mainstream business channels. In that context, Feyerabend's argument might seem a little bit irresponsible. In this context, the maturity caveat is crucial. Feyerabend's four claims represent ideals to work toward, not immediate policy proposals. Still, whether we *should* work toward them depends, in part, on how well his argument against expert authority and against the ineliminability of expert judgment fare.

10.5 Whither the Eliminability of Expert Judgment?

Considering Feyerabend's argument, should we give up on or modify our commitment to the ineliminability of expert judgment? Can we allow citizens or policymakers to decide for themselves? If we answer yes to these questions, then the arguments against the deferred decision defense of value-free science puts us in a difficult position. On the one hand, we might work to reform science such that the separability between science and policy analytically made by Mitchell can become a practical reality. If, as Feyerabend argues, we tend to grossly underestimate the competence of non-scientists to evaluate technical scientific information, then this may be the best path forward. This would allow, perhaps, for a modicum of authority to remain with science, rendered suitably value-free by being walled off from social relevance. On the other hand, if science is unavoidably value-laden, this raises significant concerns about the democratic legitimacy of science's epistemic authority. In turn, this potentially bolsters Feyerabend's claim that science is just another ideology and scientists just another interest group in society. In that case, whether or not citizens are really competent to evaluate science, they have no choice; they cannot take what scientists claim for granted.

In this section, I will explore some ideas that complement Feyerabend's approach, as well as some potential objections, in ways that I think sharpen the problem. On their basis, I will stake out two possible positions on the role of experts in society and whether we should continue to insist on the authority and eliminability of expert judgments.

10.5.1 Scientific Judgment: Transparent or Opaque?

Ian James Kidd (2016b) provides an interesting objection to Feyerabend's account of the illegitimate exercise of authority by scientists, as exemplified in the astrology case, based on Michael Polanyi's response to the case of Immanuel Velikovsky (also briefly commented on by Feyerabend, SFS 91n). Velikovsky published a book, *Worlds in Collision*, in 1950. In it, he laid out and defended a radical theory that the solar system had undergone major changes during historical times, forming the basis for many stories of ancient mythology and religion, which themselves were a source of putative evidence for Velikovsky's theory. There was an immediate and severe response from the mainstream scientific community, which attempted to censor and dismiss the book as pseudoscience. There, the scientific community showed the same sort of inappropriate dismissiveness (speaking outside of their expertise, without even really investigating the case). Polanyi argues, however, that rather than showing a lack of integrity or an illegitimate appeal to authority, their dismissiveness is entirely appropriate. Polanyi is well known for demonstrating the importance of the *tacit* dimension of scientific knowledge; the skills and implicit assumptions that make science possible. As Kidd summarized, "[w]hat seemed, to non-scientists, to be reactionary dogmatism was, in fact, a spontaneous evaluation both generated and justified by a tacit sense of plausibility. Polanyi concluded that since that sense is historically informed, collectively supported, and a product of practice and discipline, those scientists were right to trust it" (2016b, p. 476). Otherwise, they would be wasting time and resources. This points strongly toward the ineliminability of expert judgment.

A similar argument in favor of the ineliminability of expert judgment can be extracted from some ideas of Bruno Latour. Latour emphasizes not so much the dimension of tacit knowledge as the material means of science and their rhetorical function. In *Science in Action* (1987), Latour asks us to consider what is necessary to continue to dispute a scientific claim; that is, he asks us to think about the nature of arguments and counter-arguments in science. Arguments in science differ from arguments in the humanities. Arguments in the humanities are transparent – you have the textual evidence, scholarly references, the arguments, and that's it. All you need is a library card, cleverness, and motivation in order to craft a counterargument. Some aspects of the technical literature in science is, of course, like this as well – there are references to follow and logical and mathematical arguments that can be criticized. But, arguments in science are also different. Scientists construct phenomena in laboratories, which they turn into inscriptions (tables, charts, graphs, and figures), which play a special role in scientific arguments. They cannot be disputed in the way that a logical argument or an interpretation can be disputed. One may be able to visit the lab and find a flaw in the inscription device (the laboratory equipment). But, if no such flaw can be found and agreed upon, there is one further strategy available to disputants: build a laboratory of one's own and generate different results.

In the most general case in science, you cannot have a complete counter-argument without a *counter-laboratory*. This is the bread and butter of scientific argument; as Latour says, "[t]his is why all laboratories are *counter-laboratories*" (1987, p. 79). What's more, "[t]he dissenters cannot do less than the authors. ... So the dissenters do not simply have to get a laboratory; they have to get a *better* laboratory" (p. 79). This restricts effective dissent toward putative scientific facts to a very specific group with the skills and resources to operate with the proper material, technological means. Thus, while the main action of science takes place in the discursive "agonistic field" of the published literature, the anatomy of the scientific paper reveals the crucial role of the material means of the scientific laboratory. Thus, even non-scientists who are very committed, resourceful, and mature may not be able to decide for themselves without a laboratory of their own, without becoming a scientific expert in their own right.

These arguments suggest bases for scientific expertise outside the explicit (formal or informal) communications of scientists in tacit knowledge and the material means of knowledge-production; bases that cannot be completely supervised by non-experts. Ian James Kidd (2016b) offers the beginnings of a Feyerabendian response to this line of argument. As Kidd points out, the mere existence of these tacit dimensions of science does not, on its own, suffice to guarantee the public acceptance of the authority of science. If the public cannot effectively control science democratically as a result, they have to decide whether or not to trust science. For

Feyerabend, "[c]entral to his conception of the social authority of science is . . . the claim that scientists ought to conduct themselves with integrity" (Kidd 2016b, 14). Their behavior in the astrology and Velikovsky cases is reason to think that science has not yet earned that trust and authority.

10.5.2   *Authority versus Autonomy of Science*

Another challenge to Feyerabend's argument comes from recent discussions in the literature on science and values about the relationship between the authority and autonomy of science (Douglas 2009; Brown 2013). As Heather Douglas argues:

> On the basis of the value-free nature of science, one could argue for the general authoritativeness of its claims. But [given that science is not value-free] an autonomous *and* authoritative science is intolerable. For if the values that drive inquiry, either in the selection and framing of research or in the setting of burdens of proof, are inimical to the society in which the science exists, the surrounding society is forced to accept the science and its claims, with no recourse. A fully autonomous and authoritative science is too powerful, with no attendant responsibility, or so I shall argue. Critics of science attacked the most obvious aspect of this issue first: science's authority. Yet science is stunningly successful at producing accounts of the world. Critiques of science's general authority in the face of its obvious importance seem absurd. The issue that requires serious examination and reevaluation is not the authority of science, but its autonomy. (Douglas 2009, pp. 7–8)

Douglas here points out an important tension between autonomy and authority. If we think about institutions with social authority, they are or ought to be answerable to society. On the other hand, private individuals and organizations plausibly have a right to a great degree of autonomy, but they have no social authority beyond the right, which everyone has, to freely have their say and their vote. To allow a person or institution to have both would be a legitimization of autocracy.

Douglas presumes that the authority of science is not contestable, primarily on the basis of the record of its successes. If science is successful in this way, it has earned its authority. This combined with the argument that science cannot be value-free leads to the ineliminability of expert judgment and the reduction of the autonomy of science. As we have seen, Feyerabend disputes both the claim of success and the inference from success to authority. His challenge at least identifies a gap in Douglas's reasoning on this point. If authority and autonomy present the tension Douglas identifies, we appear to have the option to limit the authority *or* the autonomy of science.

On the other hand, Douglas presents an alternative option for those who share Feyerabend's concerns about the legitimacy of the social authority of science, but who are not ready to wholesale deny that authority. That authority could, potentially, be legitimized by curtailing the autonomy of science, requiring that it be guided by or answerable to the public. Call the position that Feyerabend advocates and Douglas rejects, "*epistemic anarchism.*" This position allows the functioning of society to be left alone, but radically reconfigures society's relationship to science. Call the alternative position, which preserves the authority of science but curtails its autonomy, "*strong accountability.*" This more or less retains the current dependence of society on science, or even strengthens it, but requires radical reconfiguration of the functioning of science itself. In this approach, "democratic control of science" has an even more robust meaning than in Feyerabend's consideration of that idea.

Robert Paul Wolff shares Douglas's concern about the incompatibility of authority and autonomy, though he is concerned with slightly different senses of "autonomy" and "authority" in the context of political philosophy rather than philosophy of science. Wolff is concerned about whether the moral autonomy of the individual is compatible with the legitimate authority of the state (Wolff 1970). He argues that they are not in fact compatible. He, too, considers a dilemma:

> Either we must embrace philosophical anarchism and treat all governments as non-legitimate bodies whose commands must be judged and evaluated in each instance before they are obeyed; or else, we must give up as quixotic the pursuit of autonomy in the political realm and submit ourselves (by an implicit promise) to whatever form of government appears most just and beneficent at the moment. (Wolff 1970, p. 71)

Wolff argues, on explicitly Kantian grounds, that giving up autonomy is out of the question. He thus defends *philosophical anarchism*, which denies that there is any authority over one's conduct, besides the law of morality. And, as with Kant, the moral law is something we can only give ourselves, not something that is imposed on us from outside. Feyerabend is an anarchist in precisely the same sense as Wolff, but concerning epistemic rather than ethical-political matters. He, too, is deeply committed to the autonomy of the individual. He can be taken to insist that citizens judge and evaluate knowledge claims in each instance before accepting them, rather than accepting them on some external authority.

There are thus two kinds of anarchism in Feyerabend's body of work. The one, explicitly named and frequently discussed, is known as *epistemological anarchism* (Feyerabend 1993). This thesis concerns the authority of scientific methodologies. Feyerabend argues that no single methodology, as binding, can promote the progress of science. Thus, the search by philosophers of science for a single, binding methodology is not only hopeless but also harmful. The epistemological anarchist instead recommends pluralism, contextualism, flexibility, opportunism, and creativity with respect to methodology. The broader view, which Feyerabend misleadingly termed *relativism*, is philosophical anarchism as concerned knowledge claims, which I will thus call *epistemic anarchism*. Epistemic anarchism denies any special authority over your belief or acceptance of claims, over and above what you can judge for yourself using ordinary epistemic norms, as given to yourself in the Kantian sense.

Epistemic anarchism entails the strongly curtailed authority for scientific experts that Feyerabend recommends; they will be listened to, but only as any special interest group. Epistemic anarchism does not in itself deny the value of the division of epistemic labor. What it does is deny the right of specialists within such a division to demand deference to their claims. Nor is epistemic anarchism incompatible with one's accepting some claims of scientific consensus (just as Wolff's philosophical anarchism is compatible with following some laws, if they accord with the moral law within you). Informed deliberators might decide to take that consensus as good reason to accept such claims, in some contexts, though they would be unwise to adopt this as a universal policy, as unexamined presuppositions are also sometimes the object of consensus.

Strong accountability legitimates scientific authority at the cost of its autonomy. The values incorporated into science, from the choice of research questions to the setting of burdens of proof, will need to be responsive and answerable to the public. How this could be implemented is unclear. Will scientists be able to make such value judgments on their own? Will norms of transparency be sufficient to guarantee accountability? Will new institutional structures for citizen consultation and oversight be necessary? These are the sort of questions those committed to scientific authority and the ineliminability of expert judgment need to ask.

10.6 Separation or Control?

Earlier, I identified an apparent tension in Feyerabend's thinking: should we have separation of science and state or control of science by citizens? I hastily argued that one could avoid contradiction by interpreting "control" as monitoring or evaluating rather than guiding. But perhaps it is wiser to sit with the tension. This tension reflects the other tension identified between autonomy and authority. Feyerabend was not focused on the tension in exactly the way that Douglas highlights, but the two are closely related.

Based on Feyerabend's arguments and the further considerations discussed here, I see a dilemma for the role of experts in society: On the one hand, we could strongly curtail the *authority* of scientific experts, ensure the separation of science and the state, and develop a more engaged, more mature democratic citizenry capable of adequately judging scientific results and appropriating or ignoring them as suits their needs. I have called this position *epistemic anarchism,* an extension to epistemic matters of philosophical anarchism. On the other hand, we could strongly curtail the *autonomy* of scientific practice, increase citizen oversight over or control of science, and make a radical shift of the focus of science toward the public interest. I have called this position *strong accountability.*

Each position has its benefits and drawbacks. Epistemic anarchism ensures that public values are respected by placing the moment of judgment in the public's hands or the hands of their representatives. There is no concern that private values might influence science and thus illegitimately affect our decision-making, as all scientific advice will be evaluated by nonexperts. And, it will have equal standing in public decision-making with other local, situated, indigenous, and alternative knowledges.[8] This approach fits better with a participatory democratic ideal, and it avoids the problem of illegitimate technocracy and paternalism. On the side of science, it preserves academic freedom and the right to free inquiry. It allows science to be value-neutral, if not precisely value-free. Scientists can ignore the social impact of their work without being socially irresponsible, as science would be a purely private activity.

There are, however, significant potential drawbacks of epistemic anarchism. In the political realm, history seems to show that in the absence of legitimate authority, illegitimate coercion tends to arise. Functional anarchist organizations or societies are few indeed. Inefficiencies and bad

---

[8] With apologies to Shari Clough.

decisions are likely, as ordinary individuals are prone to mistakes, (perhaps) more so than experts. Epistemic anarchism diminishes the traditional role of science as public reason. Though arguably this role has already eroded, exploding it entirely could promote further social fragmentation and disagreement. Epistemic anarchism places a heavy burden on individual citizens for education and judgment, burdens that they may not be willing to bear.

Strong accountability, by contrast, retains the intuitive authority given to scientific experts and the scientific process. It better fits with structures of representative democracy, which devolve many aspects of governance onto representatives, bureaucracies, and experts. It preserves the role of science as public reason. Perhaps its greatest benefit is that it reorients science from private interests to public-interest science. The problems of private-interest science are many, including corruption and bias (Krimsky 2003). The promise of public, mission-driven science, by contrast, is great, and its strong track record has often been underestimated (Sarewitz 2016). On the other hand, strong accountability requires that we create new norms, structures, or institutions for consultation, authorization, and regulation of science. It requires that we overcome existing institutional tensions between science and politics. And it places a significant burden on scientists to be aware of, represent, and be answerable to public values and interests. Scientists may not be willing to bear such burdens.

10.7 Conclusion: Collective Inquiry as a Third Way?

Science is unavoidably value-laden. The role of values in science cannot be deferred onto politicians, bureaucrats, or the public, but must involve the scientific experts themselves. Expert judgment is an ineliminable part of the process. So the mainstream argument for values in science goes. But Paul Feyerabend gives us reason to question the authority of scientific experts. This line of questioning leads us to a tension between the authority and the autonomy of science, and thus to a tension between two of Feyerabend's proposals: should science and the state be separate, or should there be democratic control of science? These tensions point to a dilemma in the relationship between science and society: give up on the authority of science and, with Feyerabend, become epistemic anarchists or give up on the autonomy of science and create a regime of strong accountability. My hope is that the dilemma is a false one, and we can develop a third way between these two approaches, one that retains many of the benefits and avoids many of the drawbacks of epistemic anarchism and strong account-ability. I do not have space to outline this alternative in detail, but I will gesture toward what such a view might look like.

The middle way I have in mind combines the emphases on citizen participation, the importance of science, and the value of the division of epistemic labor. It requires us to reconceive democracy along parti-cipative–democratic lines as a kind of *collective inquiry*, an idea central to the work of John Dewey (Dewey 1927; Bohman 1999). Recognizing the importance of the division of labor and the need for scientific specialists, it requires us to reconceive science-informed policy as a kind of interdisciplinary collaboration on inquiry into a shared problem. The collaboration might include scientific experts, policy-making experts, public and stakeholder representatives, and those with local, non-expert knowledge. This middle way, which I call *democracy as inquiry*, draws on various approaches that ask us to break down the traditional conceptions of science and the public and reconfigure them in ways that re-sort their traditional tasks (Dewey 1927; Bohman 1999; Latour 2004; M.B. Brown 2009, M.J. Brown 2013).

Expert authority and autonomy would both exist, with limitations, under democracy as inquiry. This approach recognizes the special authority of scientists *within* their specific area of expertise, according to the division of epistemic labor. However, this authority is not abso-lute, but rather requires situationally specific evaluation and renegotia-tion of relevance and standing in each social and policy context in which scientific results or advice are being considered. In other words, society will adopt the same trust-but-verify attitude that cautious scientists take toward each other's work. Research of primarily specialist interest, and technical decisions made within the process of basic research, would remain fairly autonomous. The limits on the social authority of such research would make this autonomy reasonably tolerable. Policymaking inquiry, in contrast, must be appropriately, publicly authorized, accoun-table, representative, and so on (see M.B. Brown 2009). Here, there must be checks in place to make sure that publicly authorized values guide inquiry. Since policy inquiry would not take past results for granted, but rather renegotiate their standing and relevance in the course of the new context of inquiry, the relative autonomy of basic research should not be a problem; it will have no immediate authority in the new inquiry. Expert judgment is ineliminable, but only one contribution to a larger process.

This is a hasty sketch of a complicated view, and I do not hope to have convinced you of its viability. But, it is one example of a way forward that might avoid some of the problems with rejecting authority or autonomy wholesale and limn the middle way between epistemic anarchism and strong accountability.

CHAPTER 11

A Way Forward for Citizen Science
Taking Advice from a Madman

Sarah M. Roe

11.1 The Citizen Science Movement

Within science, there is a movement to utilise non-scientists for an array of scientific tasks. This movement, often called citizen science, has begun gaining popularity and momentum for the better part of the last decade (Dickinson and Bonney 2012; Miller-Rushing et al. 2012; Bonney et al. 2014). The term, 'citizen scientist', first appeared in the title of a sociology book in 1995, but was quickly taken up, first within Ornithology then quickly throughout the biological sciences (Busch 2013, p. 19). But before then, there has been a long tradition of amateur practitioners making amazing scientific discoveries.[1] We need to only mention Darwin or Mendel to trigger others that have helped advance major areas of biology.[2]

Citizen scientists partake in scientific discovery, monitoring, data collection, and experimentation across a wide range of scientific disciplines. The information collected by citizens is most often used to better understand and/or predict phenomena like climate change, overexploitation, public health trends, water quality, invasive species, classification of galaxies, land use change, air quality, protein folding, pollution, and so on. Citizens are able to collect fine-grain data over regional, and sometimes continental, extents and decadal time scales, something professional scientists are unable to accomplish alone (Theobald et al. 2015). For instance, citizen science databases, like the one geared towards bird watchers, eBird, have more than 100 million records on file (Wood et al. 2011). As such, it is estimated that the scientific benefit of citizen scientists easily amounts to

---

[1] See Dawson et al. (2015) regarding the similarities between the nineteenth-century print revolution in medicine and today's citizen science movement.
[2] Feyerabend himself recognised and applauded the scientific contributions of scientists who thought of themselves as outside mainstream science, such as Einstein, Bohr and Born. See Feyerabend (1978b, pp. 88–89).

millions in in-kind economic worth (Theobald et al. 2015). It is clear citizen scientists offer increasing support for science.[3]

11.1.1   Examples of Citizen Science

Precisely because citizen science programmes are so wide-ranging in the task[4] asked of the participant, the skill level required of the participant, technology utilised for the programme, the use of the information by the scientific community and the scientific discipline targeted to utilise the data, I offer a few brief examples of citizen science to illustrate the concept.

Many birdwatchers, although not scientifically trained, provide useful scientific information, such as information regarding the location of certain species, the habits of certain birds given differing habitats and so on. The initial idea was simple: tens of thousands of people bird watch for fun every day, and this hobby can be massively beneficial for science. eBird is the world's largest biodiversity-related citizen science project, with more than 100 million bird sightings contributed each year by eBirders around the world (eBird 2018). As such, the programme has two main objectives. First, motivate the user to continually provide information regarding bird species, location of species, information about habitat, special location of the sighting and so on. Second, compile and store the information downloaded by citizens into a large usable database.

No doubt, the advancement of technology has helped in both respects. eBird is now a massive database used by many scientists and featured in many scientific studies (Wood et al. 2011). Historically, some of our current ornithological information originates from the observations of these citizen scientists. Moreover, eBird has begun implementing social features, like connecting citizen scientists with others in the area, the ability for citizens to compare their findings to others and the awarding of virtual trophies when certain accomplishments are achieved to promote use frequency (eBird 2018). Perhaps most importantly, eBird, like other social databases,

---

[3] Currently, there are at least three main organisations that support and advance the work of citizen science, namely, the Citizen Science Association, the European Citizen Science Association and the Australian Citizen Science Association. See Storksdieck et al. (2016) for an overall history of the establishment of these organisations.

[4] I am using a broad notion of 'task'. Here, tasks can range from simply letting a GPS (Global Positioning System) application confirm the location of a hiker all the way to allowing participants an active role in scientific data analysis.

has also begun emphasising education.⁵ Now, users can learn about the bird they just spotted, information about the location they are in or other possible birds in the vicinity. As a result, eBird users produce a tremendous amount of useful scientific information while learning about their area and taking part in the scientific process (Wood et al. 2011).

23andMe is a great example of citizen science used for biomedical purposes. For a minimal fee (around $150), participants can send a small sample of their own DNA that is then sequenced within a participating lab (Kelty and Panofsky 2014, p. 10). Information is then sent back to the participant, outlining the individual's disease risk, biogeographical ancestry, trait propensities and so on (23andMe 2018). Individuals are allowed, through social network functions, to share this information with one another and learn more about genetic testing, traits and disease particulars. In return, participants are asked to submit several comprehensive surveys, of which the information is used both to customise the individual's profile and add to the main database.⁶ The information collected through 23andMe is used by researchers interested in such things as population genetics, physical traits and disease like Parkinson's (Prainsack 2011).

In ecology, the current focus is on the threat of invasive alien species. In response to this concern, there has been a proliferation of volunteer-based monitoring programmes. The valuable data sets collected through these programmes facilitate large-scale, baseline population monitoring. The Invasive Plant Atlas of New England is one such regional citizen science programme, which uses trained volunteers, along with experts, to collect distribution data and detailed environmental information (Bois et al. 2011). This particular project focusses on the incorporation of true absence data,⁷ which specifically allows for the building of robust statistical models and contributes significantly to the invasive species literature (Invasive Plant

---

⁵ It may be important to note that eBird, like other citizen science projects, does indeed require some theoretical understanding from the participant. Moreover, many citizen science projects help guide and teach participants necessary theoretical knowledge. For example, eBird provides a great deal of information about habitat and facts about the species spotted, and offers other closely related species. In this way, participants are educated in the broader scientific understanding of observation and the process of speciation.

⁶ The totality of what 23andMe offers can be considered a citizen science project, as it educates the participant and offers ways in which the participant can further contribute to the database. Moreover, it connects participants with one another for further learning and processing experiences. In this way, participants are not just supplying information (DNA) to a database, but learning about themselves, such as disease processes and genetic testing, and are motivated to continue to learn and teach others.

⁷ True absence data are recorded when, repeatedly, a species is not present in a particular location. Thus, true absence data are valuable, in that it often indicates locations where the environmental conditions are unsuitable for a particular species to survive.

Atlas of New England 2018). This collaborative database allows citizen science data to be used by the public and as a data source for researchers and policymakers alike (Bois et al. 2011).

More than a million people, to date, have assisted researchers on Zooniverse, a computer platform that utilises citizens for a variety of scientific tasks. In the comfort of their own home and through the use of their own computer, citizens can study and analyse scientific findings, such as images of galaxies, historical records and videos of animals and habitats (Zooniverse 2019). Citizens are asked to answer a series of questions after learning about the subject and viewing the content or research. For example, Galaxy Zoo: Clump Scout, is an ongoing project that asks untrained citizens to help 'look' for clumpy, or cluttered, galaxies in the universe (Galaxy Zoo: Clump Scout 2019). Participants are given images of nearby galaxies to study and then asked a series of questions regarding the structure of the galaxy within the image. In this way, citizen scientists are asked to comb through images and aid researchers in mapping and tracking changes within the universe. Zooniverse offers a wide range of projects, across all sciences, in an effort to 'combine contributions from many individual volunteers, relying on a version of the "wisdom of crowds" to produce reliable and accurate data' (Zooniverse 2019).

11.1.2   The Benefits of Citizen Science

Currently, there are at least five main benefits of utilising citizens in scientific practice. First, citizen scientists contribute an astonishing amount of scientific data, as illustrated in the examples earlier. Massive databases are then utilised by scientists for their research and policymakers while drafting new policies. The importance of this contribution cannot be understated.

For example, citizen science often plays an important and indispensable role in gathering data in difficult regions. The poor accessibility of, say, alpine ecosystems makes collecting large sums of scientific data both challenging and expensive. Often alpine ecosystems are only scientifically interesting at certain times of the year, times when scientists are least likely to find the time and resources for systematic field surveys. However, the challenging season and terrain are particularly alluring for some hikers. So, while scientists are busy planning future surveys and class lectures, seasonal hikers are relied upon to continuously collect valuable data (Jackson et al. 2015).

Second, it is often argued that including citizens into the scientific process helps to educate the population regarding the goings-on in science. That is to say, citizen participation increases general awareness and knowledge among participants. Each and every citizen's science programme educates participants in some aspect of science while focussing on particular scientific phenomena. Citizens who participate in scientific activities gain knowledge about that particular area of science. So, citizens who participate via birdwatching surely gain further knowledge regarding the practice of scientific observation, the process of scientific documentation, how to utilise certain required instruments, the rarity and abundance of certain birds in a particular area, the mating and hunting rituals of certain species, and so on.

Moreover, each citizen scientist is actively engaged in the scientific process and is offered a unique way in which to experience and learn from the goings-on in science; participants will increase their understanding about the process of science through this engagement in authentic science. Similarly, one often-overlooked benefit to citizen science programmes is the promotion of an increase in scientific reasoning skills among the general population through this very process. Often seen as the most important benefit, citizen scientists can expect to receive some simple knowledge and skill sets as well as the knowledge of how to participate as a better global citizen.

Third, citizen scientists often show an increased interest in conservation, conservation practices and conservation policy, particularly if they live locally within the area. For example, one Virginia farmer stumbled upon the rare Rusty-Patched bumblebee. This lone event began a grassroots citizen science movement that culminated in many local participants, all with the intention to locate and improve the Rusty-Patched bumblebee population (Biological Conservation Newsletter 2015).[8] The hope here is that there will be a helpful feedback loop; as more citizens begin understanding how their action affects the environment and how difficult it is to reverse that effect, more people will cease the harmful action and at a much quicker rate. So, citizen science is beneficial in that it aids both science and the greater societal whole by engaging citizens, educating citizens and empowering citizen to change their actions.

---

[8] This kind of local reaction is common within the citizen science literature. Another example includes the surveying of vernal pools, which may help make local farmers more aware of conservation techniques they too can use. See Busch (2013) for more details.

Fourth, citizen science often encourages participants to participate in scientific endeavours geared towards policy and the impact scientific findings may have either locally or nationally (Busch 2013, p. 25). For example,[9] residents in Norco Louisiana began collecting air samples in the late 1990s. Concerned the local Shell Chemical plant was adversely affecting the air and causing a rise in illness within the area, residents conducted their own air monitoring using devices commonly called 'buckets'.[10] These easy to use (and make) sampling kits were utilised to collect striking evidence that the chemical plant was indeed releasing harmful levels of toxins into the air, resulting in state regulatory standards that are upheld even today (Louisiana Bucket Brigade Press Release 2000; Ottinger 2010, p. 245).[11]

Fifth, we must not forget there are also important public health advantages, in that participants that engage in fieldwork may be experiencing the outdoors or nature more frequently than other citizens. For instance, taking water inventories is an excellent excuse to spend the day kayaking in a local river or lake (Busch 2013, p. 25). Although this benefit is largely untested, there is no doubt one major motivation for citizen participation is the 'excuse' to spend more time in nature in an attempt to better understand it.

11.1.3   The Benefits Realised?

Despite many benefits, scientists are debating the value of citizen science (Cohn 2008; Elliott and Rosenburg 2019). Research shows that although citizen scientists contribute greatly to scientific data, very little of the citizen-collected research is actually being used by scientists. For instance, one study shows that while 1.3 million citizens volunteer annually, only about 12 per cent of projects geared towards citizen scientists provide data to peer-reviewed scientific articles (Theobald et al. 2015). One study indicates that scientists themselves have a hard time accepting citizen science data and contributions to the scientific process (Golumbic et al. 2017). Due to the low rate in which citizen-generated data is used within

---

[9] Another, more well-known example is Brian Wynne's work regarding sheep farmers in Cumbria Northern England and the environmental changes after Chernobyl (Wynne 1989). Wynne discusses the amount of information gathered by the sheep farmers and emphasises the need for trust in scientific knowledge and citizen scientists.

[10] See Overdevest and Mayer (2008) for more information regarding the materials needed to make these simple collection units and the use of 'buckets'.

[11] The Louisiana Bucket Brigade has now extended its mission to end all petrochemical pollution in Louisiana. For current information and findings collected by the group, please visit their website at www.labucketbrigade.org.

science, the ever-growing citizen science movement is only impacting scientific knowledge in a small and non-proportionate way.

Typical concerns include wider variability among data collected by citizen scientists (Genet and Sargent 2003; Case et al. 2016; Ellwood et al. 2016), over and underestimations among data collected by citizen scientists (Bray and Schramm 2001), misidentification among data collected by citizen scientists (Genet and Sargent 2003), lack of useful information submitted along with data collected by citizen scientists and so on (Jackson et al. 2015). That is, studies are finding that scientists perceive citizen-collected data as only useful within or pertaining to certain sciences, for example, biological sciences and conservation sciences, and don't believe citizens' participation could yield the type of information needed in other fields such as chemistry and physics (Motion 2019). Still more scientists are weary of data collected by citizens due to the belief that untrained observers without proper scientific context will overestimate the data. So, asking an untrained observer to look for something specific creates an increased chance of the observer, due to a mixture of under-training and overenthusiasm, to 'observe' more of the objects than there actually are. The very real fear that citizen-collected data are skewed, for one reason or another, prevents some scientists from utilising citizen-generated databases.

Advanced statistical tools are continuously being developed to help determine and correct potential bias in citizen science databases and determine how citizen-collected data is best used in scientific investigation (Theobald et al. 2015). Moreover, comparisons between citizen science and expert-driven data are on the rise in the scientific literature (Muenich et al. 2016; Wilderman and Monismith 2016). One study, while utilising five statistical modelling techniques, compared two databases, one obtained by scientists or experts and the other utilisings opportunistic data or data obtained by citizen scientists. The study found little difference between the expert and citizen science databases (Jackson et al. 2015). Other studies argue for the importance of standard setting[12] both for the development of citizen science projects and the use of citizen science-collected information in the scientific literature (Ottinger 2010). Still other studies analyse the recent efforts within citizen science projects to increase the credibility of

---

[12] For example, the EPA (US Environmental Protection Agency) offers a handbook and templates for those interested in coordinating a citizen science project. See United States Environmental Protection Agency (2019).

the collected data (Freitag, Meyer, and Whiteman 2016; Muenich et al. 2016).

From the abovementioned examples, it becomes clear that citizen scientists can play an important role in predominantly the accumulation of data.[13] However, archiving and analysing the large sets of data produced by citizen scientists is completed by scientists (Kolok 2011, pp. 629–630). While simple tools and pervasive technology can be handed off to citizen scientists rather easily during short training sessions,[14] large data sets are then produced, monitored and archived by scientists. Moreover, experts or scientists can then use follow-up or confirmatory mechanisms to check the data gathered by citizen scientists (Wood 2011, p. 3). Kelty and Panofsky (2014) found that while most current citizen science programmes promote and produce educated citizen scientists, very few consider allowing participants/citizen scientists in agenda setting or applying theory to the data gathered. Other studies show that although particularised knowledge is gained during citizen science training, often knowledge regarding scientific methods and practices, application of knowledge to the broader world, change of behaviours in response to gained knowledge and so on, was not significantly affected (Trumbull et al. 2000; Trumbull et al. 2005; Jordan et al. 2011; Kelty and Panofsky 2014). Another study found that intensive training programmes for citizens do increase particularised knowledge, such as knowledge regarding invasive species and location of particular plant species but do little to increase citizen awareness of scientific methods or processes (Jordan et al. 2011, pp. 1152–1154).

That is to say, contemporary citizen science programmes often do not allow for citizen participation in a widespread way. Thus, the role of the citizen scientist is, more often than not,[15] seen as the first tier of the scientific process, one of scientific data gathering.[16] From these findings,

[13] It may be important to distinguish between citizen science projects that focus citizen efforts on the collection of data (e.g. eBird) and those that are geared towards effecting social change (e.g. Norco's 'Bucket Brigade'). This distinction will be discussed later in the chapter. For more information regarding differing citizen science projects and the terminology utilised within the field, see Eitzel et al. (2016).

[14] For example, volunteer citizen scientists are currently being utilised to mine the vast biomedical literature for important information required by specific researchers. This task requires that the citizen scientist learn and utilise computer mining technology (Tsueng et al. 2016).

[15] Of course, there are examples of contemporary citizen science programmes that do offer a more comprehensive scientific experience to the participants. These will be discussed, and lauded, later within the article. For now, I only wish to draw the reader's attention to the mass use of citizen scientists for data collection. This is particularly true within the physical sciences and chemical sciences.

[16] It should be noted, Paul Tibbetts argues similarly when he points out Feyerabend's shaman and witchdoctor can make initial discoveries but are not in the right position to exploit their discovery

it would seem as though the citizen science movement has not reached its full potential. While primarily focussing on the use of citizens to collect much-needed and useful data, the movement has yet to properly and ubiquitously emphasise the other three benefits, namely education of the participant, increased participant interest in conservation and policy, and increased engagement in scientific and societal concerns. For one way forward, we turn to Paul Feyerabend, the madman of science.

11.2 Taking Advice from Paul Feyerabend

Described in *Nature* as the 'worst enemy of science' (Theocharis and Psimopoulos 1987), Feyerabend's writings have often invoked strong reactions from his readers.[17] Upon first glance, it would seem as though Feyerabend rejects all rational standards, any discrimination between ideas and pushes for relativism (Kulka 1977, p. 277). For example, the slogan 'anything goes' is first used by Feyerabend within an address delivered at Loyola University in 1970, and subsequently appears in *Against Method* at the end of chapter 1 (Feyerabend 1975a, p. 19). Since that time, philosophical literature has been peppered with criticism and cries of outrage towards Feyerabend's call for anarchy. Many have speculated on what exactly was meant by the slogan,[18] and even more philosophers and scientists have quickly discarded Feyerabend's antidote as the obvious ramblings of a madman.[19]

Alternately, others have argued that we should regard Paul Feyerabend's work with much less anxiety (Hacking 1991; Naess 1991; Churchland 1997).[20] Instead of thinking of Feyerabend as the madman of science, it is much more fruitful to understand his work as pluralistic, progressive and fluid.[21] Feyerabend's work is a call for an openness with respect to different views

---

(Feyerabend 1978b, p. 56). For Tibbetts, Feyerabend greatly emphasised the role of non-scientists in science by not acknowledging their shortcomings within other parts of the scientific process.

[17] Among the most famous negative reactions are Joseph Agassi (in Munévar 1991, pp. 379–387) and Herbert Schadelback (in Munévar 1991, pp. 433–439).

[18] See Oberheim (2006) and Shaw (2017) for more details.

[19] See Nagel (1977); Worrall (1978); Laudan (1996); Godfrey-Smith (2003, p. 117); Bernstein (2011); Agassi (2014). Also, Shaw (2017) for examples and further discussion.

[20] For example, Eric Oberheim argues that despite Feyerabend's playful and proactive manner of writing, a reader must simply pay closer attention to the details of his argument to find a better appreciation of Feyerabend's work (Oberheim 2006). As such, many of the once confounding statements and alleged ambiguities resulted from a quick read or a quick reaction from the reader.

[21] See Naess (1975); Kulka (1977); Tibbetts (1977); Meynell (1978); Zahar (1982); Bailin (1990); Maia Neto (1991); Munévar (1991); Sankey (1994); Kresge (1996); Lloyd (1996); Lloyd (1997); Preston (1997a); Preston (2000); Munévar and Lamb (2000); Munévar (2000); Oberheim and Hoyningen-Huene (2000); Geelan (2001); Farrell (2003); Oberheim (2006); Roe (2009); Kidd (2011); Meyer (2011); Kidd (2013); Bschir (2015).

and the acceptance of plurality.[22] Feyerabend envisioned science as a marketplace of ideas in which many options can be explored (Shaw 2018b). As a result, Feyerabend argued that all individuals have something to add to science; even the ramblings of madmen have something to offer science (Feyerabend 1975a, p. 68).

Although Feyerabend's idyllic scientific community has yet to be realised, citizen science may bring us one step closer. To better understand a possible way forward for the citizen science movement, we shall focus on two of Paul Feyerabend's important insights, namely, the scientific expert and the role of citizens as a counterbalance to that scientific expertise.

11.2.1   The Scientific Expert

In *Farewell to Reason*, Feyerabend contemplates two definitions of experts. First, 'an expert is a person who produces important knowledge and has important skills. His knowledge and skill must not be questioned or changed by non-experts' (Feyerabend 1987, p. 55). For, Feyerabend, education often educates students in just one way of thought. This focuses the student and provides expertise in one area. He argues, 'From the very beginning, intellectuals claimed to possess insights unavailable to ordinary mortals' (p. 115). According to this definition, experts are people with specialised skills, that is, skills that limit their ability to be open and perceptive to all different parts of nature. Perhaps most importantly, this definition grants experts a superior knowledge that cannot be challenged.

By thinking of scientists as this type of expert, not only do we get a very narrow indication of what nature looks like, we get ideas about nature that ought not to be challenged. For Feyerabend, this was disastrous, as one of the worst things a society can do is 'grant experts special privileges' (Feyerabend 1975a, p. 37). If left unchecked,[23] Feyerabend believed experts 'have the *power* to enforce their wishes, and because they *use* this power just as their ancestors used *their* power to force Christianity on the peoples they encountered during their conquests' (p. 299). Most importantly, this type of unchecked power would be detrimental to science, as there is little room for imagination, creativity, risk-taking or a plurality of theories and methods (p. 38). Indeed, Feyerabend argues that this type of narrowed knowledge cannot be trusted, even from experts (Feyerabend 1970b, p. 390).

---

[22] See Tibbetts (1977); Sankey (1994); Lloyd (1996); Preston (2000); Farrell (2003); Roe (2009).
[23] Some argue that Feyerabend only meant to show that science is a far more complex entity than was previously thought, and that science is only tyrannical if we isolate science from society. See Kidd (2011).

The second definition points out that 'experts in arriving at their results often restrict their vision. They do not study all phenomena but only those in a special field; and they do not examine all aspects of those special phenomena but only those related to their occasionally rather narrow interests' (Feyerabend 1987, p. 56).[24] Like the first definition, this expert focuses her training and skills to understand nature from one very limited perspective. However, this type of expert understands her limits and boundaries and that her ideas are not gifted an exalted status. Feyerabend seems to favour this definition of a scientific expert. Under this definition, scientific experts, while understanding their limitations, narrowed interests and specialised training, recognise that the creation of scientific knowledge may require additional methods, subsequent modification, particular training or technology, or novel and risky ideas.

As such and according to Feyerabend, scientific experts should be unscrupulous opportunists who are not afraid to work outside the rigid guidelines set by the scientific discipline (Feyerabend 1975a, p. 10). For Feyerabend, the scientific community should be made up of skilled individuals who recognise their limits, are open to constructive criticism and are willing to utilise any method or theory necessary. Feyerabend asserts that scientists should be opposed to all systems of rules and constraints, willing and able to think outside of their training, opportunistic and creative (Roe 2009, p. 14). Indeed, this disposition may be both the very antidote for modern science and a way forward for the citizen science movement.

11.2.2   The Citizen as a Counterbalance to Scientific Expertise

Second, Feyerabend argues that an important part of science is participation in science. He states, 'The knowledge we need to understand and to advance the sciences does not come from theories, *it comes from participation*' (FtR, p. 284 emphasis added). Although strongly worded, it seems that Feyerabend's insight does allow more space for the citizen scientist, in that Feyerabend emphasises, not training or specialty, but the practice of or participation in the activity of science. Indeed, this seems to mirror the very spirit of the citizen science movement, namely, the active participation of untrained citizens in the scientific process for the mutual benefit of both the citizen and science. Feyerabend argues that 'research is not a privilege of special groups and (scientific) knowledge not a universal measure of

---

[24] A very similar definition can be found in Feyerabend (1970, p. 389).

human excellence' (pp. 27–28). Instead, Feyerabend may provide a more active role for citizens within scientific practice.

In particular, Feyerabend may offer a better way to understand the role of the citizen within science when he argues that science needs a counterbalance, or an outside perspective (Feyerabend 1975a, pp. 52–64). For Feyerabend, citizens 'are not tied to any particular method or world-view' (p. 247), and, as such, may be in a better position, at times, to help guide science (p. 43). Each individual, expert and citizen alike, engaging in science will have a different and important perspective, something to add to scientific knowledge (p. 215). Indeed, the citizen, because of his or her different perspective may add to scientific practice in a number of very helpful ways[25] (Feyerabend 1987, pp. 58–60).

In his view, 'Citizens, guided but not replaced by experts, can pinpoint . . . short comings' within science (FtR, p. 57). That is to say, citizens may be able to provide science with a different and necessary perspective, one science would not have without citizen input. Feyerabend writes:

> The objection that citizens do not have the expertise to judge scientific matters overlooks that important problems often lie across the boundaries of various sciences so that scientists within these sciences don't have the needed expertise either . . . But the competence of the general public could be vastly improved by an education that exposes expert fallibility instead of acting as if it didn't exist. (Feyerabend 1975a, p. 251)

So, it may be the case that citizen-gathered data can bring into focus a new theory, citizen input may uncover problems in theories, or citizen input may show one theory to be more useful than another, thus aiding the scientific process in precisely the manner Feyerabend believed (Feyerabend 1975a, pp. 106–122). In this way, an increase in scientific diversity (ibid) and adherence to the idea that no standpoint is better than any other (Feyerabend 1991, p. 503) provides the basis for a more active role for the citizen within science.

Feyerabend's insight reaches beyond the role of the citizen scientist, as his opportunistic scientist should also welcome both data gathered by and input offered by the citizen scientist.[26] In his own words, 'Successful

---

[25] Particular ways in which the citizen scientist may aid in the scientific process will be discussed later in the chapter.

[26] Of course, this assumes that the research mentioned earlier, comparing the data collected by citizens and data collected by experts, is correct and there is very little difference between the two data sets; differences that would cause issue if used during the scientific process. If there is a difference, or if error is found within the citizen-collected data, then even Feyerabend's opportunistic scientist should question using such data, but this would require input and conversion between both the expert and the citizens.

research does not obey general standards; it relies now on one trick, now on another, and the moves that advance it are not always known to the movers' (Feyerabend 1987, p. 281). So, citizen-driven data collection would be considered yet another trick that may aid the scientist at any given time. Additionally, citizen-offered input on methodology or structure of the scientific experimentation and so on would be considered further tricks that may aid the scientist at any given time. As a result, citizen scientists can be very useful for the modern scientist, in that 'a scientist who wishes to maximize the empirical content of the views he holds and who wants to understand them ... must adopt a pluralistic methodology. He must compare ideas with other ideas' (Feyerabend 1975a, 30). However, the scientist must be open to this new diversity within science and accepting of the idea that citizens can offer valuable and reliable insight into the scientific process and within scientific practice.

Feyerabend saw a society in which science and citizens were not divided but worked hand in hand when he writes, 'We arrive at the result that the separation of science and non-science is not only artificial, but detrimental to the advancement of knowledge' (Feyerabend 1975a, p. 173). A variety of opinion is necessary for objective knowledge and this variety is dependent on an influx of views that are associated with differing world-views, training, technology, values, and so on. (p. 46). As such, Feyerabend argues for the abolishment of hierarchy, in that we must not treat experts as having a special knack for reality, truth and knowledge. Instead, Feyerabend argues for the rightful place of citizens in science and the crafting of knowledge Feyerabend 1987.

11.3 Science and Society: A New Relationship

This chapter began by identifying the five main benefits of utilising citizens in scientific practice, namely, mass data collection, education of the participant, increased participant interest in conservation and policy, increased engagement in scientific and societal concerns, and increased public health. Currently, we find that the citizen science movement has primarily focussed on the first benefit, in that citizen scientists surely do contribute an astonishing amount of scientific data that is then utilised by scientists, policymakers, and so on.[27] From the insights offered previously, it strikes me that Feyerabend would dislike the narrow and disengaged way in which

---

[27] It may be important to distinguish between citizen science projects that focus citizen efforts on the collection of data (e.g. eBird) and those that are geared towards effecting social change (e.g. Norco's 'Bucket Brigade'). Many thanks to Kevin Elliott (personal correspondence) for pointing out this important distinction.

science is currently utilising citizens and their wealth of talent. Although contemporary citizen science programmes do often take advantage of skills already acquired by citizens, far more can be done. Feyerabend would champion citizen science on a more multi-levelled or widespread way, where the benefits of citizens could be properly amplified within the sciences. Once we rid ourselves of the idea that only scientists have ideas and special methods for improving those ideas that generate ultimate knowledge, we can begin to clear a space for a myriad of non-expert individuals into science (Feyerabend 1969c, p. 791; 1975d, p. 167). In other words, Feyerabend would champion the complete admixture of expert and citizen precisely because he thought citizens had far more to offer science.

As such, Feyerabend would champion the infiltration of citizens at every level in science. It is only through this process can science truly progress, in that it allows for a plurality of methods, theories, ideas, viewpoints and talents. That is to say, Feyerabend would push the citizen science movement into a more radical direction, one that moves beyond mere data collection. Citizen science projects should emphasise all key components to the scientific pursuit, both for the advancement of science as well as the betterment of the citizens. Information regarding scientific methods, theory construction, experimental procedures, policy awareness and crafting, environmental concerns, behavioural changes in response to the information, and so on should all have a place with citizen science-based projects. Only in this way, can citizens inform science in the manner argued by Feyerabend, as one of counterbalance and as part of the marketplace of ideas and talents.

Please note, I am not alone when arguing for a more radical direction for citizen science-based projects. Sandra Henderson, director of the citizen science project BudBurst, argues similarly, in that 'Scientists and educators must also work to better engage [citizen science] participants beyond data collection and to immerse them more fully in the scientific process, including study development, interpretation, and reporting of results' (Henderson 2012, p. 283). Similarly, some are arguing for even more use of citizen-gathered data by showing previous concerns regarding variability, underestimations and misidentifications of citizen-collected data are vastly

---

While it is true that citizen science programmes that aim at social change or begin locally as a grass roots programme do tend to incorporate the citizen throughout the scientific process in a more through way, it is still not the case that these types of projects meet all five criteria. That is to say, citizen science projects geared towards effecting social change often do a better job at incorporating citizens into the scientific process, but still more can be done.

overestimated and thus can be overcome (Ottinger 2010; Jackson et al. 2015; Theobald et al. 2015; Elliott and Rosenberg 2019). However, I believe, as argued earlier, Feyerabend provides us with several key insights that, when embraced, offer a particular citizen-centred approach to the citizen science movement, an approach that may very well lead to the achievement of all five goals of the movement.

11.3.1 Examples of Citizen-Focussed Citizen Science

One way forward for the citizen science movement is to embrace and incorporate important perspectives that have, at times, been neglected by the scientific community. For example, indigenous peoples and their knowledge of the management of global ecosystems has not gone unnoticed by scientists. Information regarding threatened species, invasive species, aquatic ecosystems and climate change can be offered by indigenous populations and their documentation. That is, the potential contribution of indigenous knowledge to contemporary ecosystem science and management is irrefutable. However, use and documentation of use of indigenous biocultural knowledge is often lacking. One team found that more of this valuable information was being used after the 1970s, but still numbers indicate the low use of this invaluable research by scientists (Ens et al. 2015). Indeed, engagement of indigenous people in large conservation agendas may promote more holistic socio-economic thinking among scientists and citizens, as well as greater progress towards traditional scientific values, such as data collection and broad knowledge.

Alison Wylie argues similarly when she promotes 'intellectual as well as pragmatic partnerships with descendant communities, especially Aboriginal and Indigenous communities' within the discipline of archaeology (Wylie 2015, p. 189). These partnerships are more than typical scientific collaborations, in that indigenous peoples play an equally robust role alongside scientists. Crafting research agendas, challenging concepts and methodologies are all meaningful and important ways in which citizen scientists can add to the scientific endeavour (Wylie 2014; Wylie 2015).

Another way forward for the citizen science movement is to actively utilise data gathered by citizen scientists and modify it in ways that are unprecedented. One team of scientists estimated the population size of golden eagles in Pennsylvania by utilising citizen-gathered raw data and advanced models that account for animal movement and migration in combination. The result is a more accurate population estimate, one that would not have been achieved without large data sets from citizens covering more terrain than traditional academic or scientific data sets or the model produced by scientists and citizens to understand animal movements. In combination, scientists and citizens provide a more accurate estimation of the golden eagle population (Dennhardt et al. 2015).

Another way forward for the citizen science movement is to trust in the abilities of participating citizens by recognising that worries regarding citizen scientists' abilities to learn new techniques, skills and technology are unfounded. For instance, it is difficult to estimate real costs to geographically diffuse environmental pollution. To better understand the widespread degradation of human-produced pollution, massive amounts of data across a spectrum of geographical locations are necessary. Luckily, citizen scientists can now monitor biologically active compounds in local environments easily and safely. Scientific supply companies are currently producing strips that allow for the testing of environmentally relevant concentrations of various heavy metals and pollutants in water (Kolok 2011). These kinds of water chemical assay test, frequently thought to be too complicated for average citizens to manoeuvre, are very similar to pregnancy tests and pool chemical strips used by millions of people worldwide annually. Citizen scientists are indeed capable of learning new technologies or managing new skills once taught how to do so.

Another way forward for the citizen science movement is to embrace and incorporate citizen scientists at all levels within scientific practice. For instance, Sharon and Patrick Terry founded PXE International, a patient advocacy group representing those with a rare genetic disorder called pseudoxanthoma elasticum (PXE). PXE International advocates citizen scientists to participate in research as active and critical collaborators as well as donors of rare genetic material. Though lacking formal scientific training, the Terrys participated actively in the laboratory research that lead to the discovery of the mutation responsible for PXE. The Terrys, through their organisation PXE International, founded a blood and tissue bank for researchers of PXE, collaborated closely with researchers around the world who work on PXE, and presented papers and posters at scientific conferences on the topic of PXE (Terry and Boyd 2001, pp. 181–182). Perhaps most notably, the Terrys were in constant contact with researches, often daily, regarding findings and progress. The incorporation of motivated, passionate citizens into areas like laboratory work and research would benefit not only scientific progress, but also society as a greater whole.

Another way forward for the citizen science movement is to champion the use of scientific information by the public in unique ways, perhaps even interpreting data and analysing theories. For instance, SNPedia is an online wiki that allows users to interpret their own genetic single nucleotide polymorphisms (SNPs) by allowing users to make their own evaluations and pursue information about the actual molecular biology in the scientific literature (Cariaso and Lennon 2012). In this way, citizens are asked to not only add to or collect data but also make comments about the data or interpret the data as they understand it. Thus, this form of citizen science solicits non-expert, non-mainstream understandings of genetic clustering, and asks participants to explore new ideas regarding genetic theory. This type of citizen science programme allows citizens to take a more active role in science, theory and interpretation of data.

Another way forward for the citizen science movement is to convert scientific information into materials that are suitable for the public. These materials aid in furthering public scientific knowledge and are often found in accessible locations such as the internet, newspapers and magazines, all in an effort to reach the ever-growing population of citizen scientists. For example, information regarding possibly harmful compounds in plastic water bottles has changed individual choices regarding the type of bottles used and has led to shifts in the personal beverage container market (Kolok 2011, p. 627). By making scientific information more user-friendly and easier to access, complex scientific findings can be modified to impact both the public and citizen scientists. In this way, citizen science materials are reaching a broader audience, educating continuing citizen scientists more rapidly and ensuring the continuation and participation of the citizen science movement.

Another way forward for the citizen science movement is to build citizen-based projects around the needs of the citizen. One good historical example is the AIDS epidemic, in which citizen scientists played an important role in furthering medical research. Epstein's work shows that public understanding of complex science is achievable as long as there is motivation, such as in cases of life-threatening disease. And, public understanding can be turned into public participation in the research process, funding and advocacy for further medical research (Epstein 1996).

For these reasons, some citizen science projects have come to realise the importance of citizen scientist retention and motivation. Some studies are underway to pinpoint ways in which project designers can achieve citizen science success, retention and motivation (Becker-Klein, Peterman and Stylinski 2016; West and Pateman 2016; Frensley et al. 2017; Seymour and Haklay 2017). By building the science project around the participants, recognising that both scientists and citizens are part of the project, all parts of the scientific process are more accessible to the participants.

11.3.2   Science for the Citizen

By understanding the contemporary citizen science movement alongside Feyerabend's views, the way forward becomes clearer; in order for science and society to benefit in the ways the citizen science movement hopes, more integration between citizens and science must occur, and importantly, at *every level*. Feyerabend taught us that while the current citizen science movement is primarily focussed on what the citizen can do for science and what the citizen can learn from science, the movement, moving forward, should also focus on what science can do for the citizen and what science can learn from the citizen. I hope the insights offered previously provide the reader with a better understanding of why the citizen science movement must promote scientific education and a broader knowledge to participants, increase citizen interest in conservation and policy, and increase both citizen local and national engagement, while promoting a rewarding experience for both the expert and citizen alike. By allowing citizens into science at every level, all five[28] benefits are reached, which truly result in *science in a free society*.[29]

---

[28] The five benefits referred to here are those outlined earlier in the chapter regarding the benefits of citizen science. However, Feyerabend would argue that many more benefits will arise from this type of partnership, such as more robust scientific knowledge and democratic inclusivity. Exuberant thanks to Jamie Shaw for pointing this out during personal communication.

[29] Acknowledgements: I am endlessly appreciative to Jamie Shaw and Karim Bschir for inviting me to contribute to this volume and for all their helpful feedback and encouragement along the way. This chapter has been presented at a number of philosophical and scientific conferences, such as the American Association for the Advancement of Science-Pacific Division, the International Society for the History, Philosophy and Social Studies of Biology, the Philosophy of Science Association, the Society for Philosophy of Science in Practice, and the Center for Values in Medicine, Science, and Technology. The advice and guidance received at these conferences has helped to shape and strengthen this chapter. I am particularly grateful to both Ian James Kidd and Matthew J. Brown who have provided both helpful feedback and support for this project. A very special thanks to Roberta Millstein who continues to support me and my work in every way. Finally, I am grateful for David J. Buller who has continued to support my Feyerabend obsession.

Bibliography

23andMe. 2018. "DNA Reports List." Online www.23andme.com/dna-reports-list/. Accessed date: September 3, 2019.

Achinstein, P. 1993. "How to Defend a Theory without Testing It: Niels Bohr and the 'Logic of Pursuit'." *Midwest Studies in Philosophy* 13: 90–120.

Agassi, J. 2014. *Popper and His Popular Critics: Thomas Kuhn, Paul Feyerabend, and Imre Lakatos.* New York: Springer.

Ambrosio, C. 2015. "Objectivity and Representative Practices across Artistic and Scientific Visualisation." In *Visualisation in the Age of Computerisation*, A. Carusi, A. Hoel, T. Webmoor, and S. Woolgar (eds.), New York: Routledge, 118–144.

Anderson, E. 1995. "Knowledge, Human Interests, and Objectivity in Feminist Epistemology." *Philosophical Topics* 23: 27–58.

Aristotle and Halliwell, S. 1987. *The Poetics of Aristotle: Translation and Commentary.* Chapel Hill: University of North Carolina Press.

Bailin, S. 1990. "Creativity, Discovery, and Science Education: Kuhn and Feyerabend Revisited." *Interchange* 21(3): 34–44.

Barseghyan, H. 2015. *The Laws of Scientific Change.* New York: Springer.

Barseghyan, H. 2018. "Redrafting the Ontology of Scientific Change." *Scientonomy* 2: 13–38.

Barseghyan, H. and Mirkin, M. 2019. "The Role of Technological Knowledge in Scientific Change." In M. Héder and E. Nádasi (eds.), *Essays in Post-Critical Philosophy of Technology.* Wilmington: Vernon Press, 5–17.

Barseghyan, H. and Shaw, J. 2017. "How Can a Taxonomy of Stances Help Clarify Classical Debates on Scientific Change?" *Philosophies* 2(4): 24.

Becker-Klein, R., Peterman, K., and Stylinski, C. 2016. "Embedded Assessment as an Essential Method for Understanding Public Engagement in Citizen Science." *Citizen Science: Theory and Practice* 1(1): 1–6.

Bell, J. 1964. "On the Einstein Podolsky Rosen Paradox." *Physics* 1(3): 195–290.

Bernstein, R. 2011. *Beyond Objectivism and Relativism: Science, Hermeneutics, and Praxis.* Philadelphia: University of Pennsylvania Press.

Betz, G. 2013. "In Defence of the Value Free Ideal." *European Journal for Philosophy of Science* 3(2): 207–220.

Biagioli M. 1993. *Galileo, Courtier: The Practice of Science in the Culture of Absolutism.* Chicago, Ill.: University of Chicago Press.

Biagioli. M. 2006. *Galileo's Instruments of Credit: Telescopes, Images, Secrecy.* Chicago, Ill.: University of Chicago Press.

Bickle, J. 1998. *Psychoneural Reduction: The New Wave.* Cambridge, MA: MIT Press.

Biological Conservation Newsletter. 2015. "Rare Rusty-Patched Bumblebee Discovered in Virginia Survey." Smithsonian National Museum of Natural History, 361. Retrieved from https://insider.si.edu/2014/10/rusty-patched-bum ble-bee-discovered-smithsonian-researchers-find-rare-bee-thought-headed-exti nction/. Last date accessed: November 18, 2020.

Boghossian, P. 2006. *Fear of Knowledge: Against Relativism and Constructivism.* Oxford: Clarendon Press.

Boghossian, P. 2011. "Three Kinds of Relativism." In *A Companion to Relativism,* S. D. Hales (ed.), Malden, MA: Wiley-Blackwell, 53–69.

Bohm, D. 1952. "A Suggested Interpretation of the Quantum Theory in Terms of 'Hidden' Variables. I." *Physical Review* 85(2): 166–179, 180–193.

Bohm, D. 1953. "Proof That Probability Density Approaches $|\psi|^2$ in Causal Interpretation of the Quantum Theory." *Physical Review* 89(2): 458–466.

Bohm, D. 1957a. "A Proposed Explanation of Quantum Theory in Terms of Hidden Variables at a Sub-Quantum-Mechanical Level." In *Observation and Interpretation: A Symposium of Philosophers and Physicists, Proceedings of the Ninth Symposium of the Colston Research Society,* S. Körner and M. Pryce (eds.), London and New York: Butterworths Scientific Publications, 33–40.

Bohm, D. 1957b. *Causality and Chance in Modern Physics.* London: Routledge & Kegan Paul.

Bohm, D. and J. Vigier. 1954. "Model of the Causal Interpretation of Quantum Theory in Terms of a Fluid with Irregular Fluctuations." *Physical Review* 96(1): 208–216.

Bohm, D. and Y. Aharonov. 1957. "Discussion of Experimental Proof for the Paradox of Einstein, Rosen, and Podolsky." *Physical Review* 108(4): 1070–1076.

Bohman, J. 1999. "Democracy as Inquiry, Inquiry as Democratic: Pragmatism, Social Science, and the Cognitive Division of Labour." *American Journal of Political Science* 43(2): 590–607.

Bois, S., Silander Jr., J. , and Mehrhoff, L. 2011. "Invasive Plant Atlas of New England: the Role of Citizens in the Science of Invasive Alien Species Detection." *BioScience* 61(10): 763–770.

Bonney, R., Shirk, J., Phillips, T., Wiggins, A., and Ballard, H. 2014. "Next Steps for Citizen Science." *Science* 343: 1436–1437.

Brahe, T. 1598/1946. *Tycho Brahe's Description of His Instruments and Scientific Work as given in Astronomiae Instauratae Mechanica,* H. Ræder, E. Strömgren, and B. Strömgren (eds.) (trans.), København: Det Kongelige Danske Videnskabernes Selskab.

Brannigan, A. 1981. *The Social Basis of Scientific Discovery.* Cambridge: Cambridge University Press.

Bray G. and Schramm Jr., H. 2001. "Evaluation of Statewide Volunteer Angler Diary Programme for Use of a Fishery Assessment Tool." *North American Journal of Fishing Management* 21: 606–615.

Brentano, M. 1991. "Letter to an Anti-Liberal Liberal." In *Beyond Reason: Essays on the Philosophy of Paul Feyerabend*, G. Munévar (ed.), Boston: Kluwer Academic Publishers, 199–212.

Bright, L. 2018. "Du Bois' Democratic Defence of the Value Free Ideal." *Synthese* 195(5): 2227–2245.

Brown, J. 2001. *Who Rules in Science?* Cambridge: Harvard University Press.

Brown, J. 2002. "Funding, Objectivity and the Socialization of Medical Research." *Science and Engineering Ethics* 8(3): 295–308.

Brown, J. 2008a. "Politics, Method, and Medical Research." *Philosophy of Science* 75(5): 756–766.

Brown, J. 2008b. "The Community of Science." In *The Challenge of the Social and the Pressure of Practice: Science and Values Revisited*, Martin Carrier, Don Howard, and Janet A. Kourany (eds.), Pittsburgh, PA: University of Pittsburgh Press, 189–216.

Brown, J. 2016. "Patents and Progress." *Perspectives on Science* 24(5): 505–528.

Brown, M. B. 2009. *Science in Democracy: Expertise, Institutions, and Representation.* Cambridge, MA: MIT Press.

Brown, M. J. 2009. "Models and Perspectives on Stage: Remarks on Giere's *Scientific Perspectivism*." *Studies in History and Philosophy of Science (Part A)* 40: 213–220.

Brown, M. J. 2013. "The Democratic Control of the Scientific Control of Democracy." In *EPSA11 Perspectives and Foundational Problems in Philosophy of Science*, V. Karakostas and D. Dieks (eds.), Dordrecht: Springer, 479–492.

Brown, M. J. 2016. "The Abundant World: Paul Feyerabend's Metaphysics of Science." *Studies in History and Philosophy of Science* 57: 142–154.

Brown, M. J. 2020. *Science and Moral Imagination: A New Ideal for Values in Science.* Pittsburgh: University of Pittsburgh Press.

Brown, M. J. and Kidd, I. J. 2016. "Introduction: Reappraising Paul Feyerabend." *Studies in History and Philosophy of Science* 57: 1–8.

Bschir, K. 2015. "Feyerabend and Popper on Theory Proliferation and Anomaly Import: On the Compatibility of Theoretical Pluralism and Critical Rationalism." *HOPOS: The Journal of the International Society for the History of Philosophy of Science* 5(1): 24–55.

Bunge, M. (ed.) 1967. *Quantum Theory and Reality: Studies in the Foundations, Methodology and Philosophy of Science.* Berlin: Springer.

Burian, R. 1971. *Scientific Realism, Commensurability, and Conceptual Change: A Critique of Paul Feyerabend's Philosophy of Science.* Pittsburgh: University of Pittsburgh, Ph.D. dissertation.

Busch, A. 2013. *The Incidental Steward: Reflections on Citizen Science.* New Haven: Yale University Press.

Cariaso, M and G. Lennon. 2012. "SNPedia: a Wiki Supporting Personal Genome Annotation, Interpretation, and Analysis." *Nucleic Acids Research* 40: D1308–D1312.

Carnap, R. 1935. *Philosophy and Logical Syntax*. London: Routledge.

Carnap, R. 1950. "Empiricism, Semantics, and Ontology." *Revue Internationale de Philosophie* 4(11): 20–40.

Carter, B. 1971. "Causal Structure in Spacetime." *General Relativity and Gravitation* 1: 349–391.

Case, N., MacDonald, E. A., McCloat, S., Lalone, N. and Tapia, A. 2016. "Determining the Accuracy of Crowdsourced Tweet Verification for Auroral Research." *Citizen Science: Theory and Practice* 1(2): 1–9.

Chalmers, D. 1995. "Facing Up to the Problem of Consciousness." *Journal of Consciousness Studies* 2(3): 200–219.

Chang, H. 2004. *Inventing Temperature: Measurement and Scientific Progress*. Oxford: Oxford University Press.

Chang, H. 2012. *Is Water H_2O?: Evidence, Realism and Pluralism*. Dordrecht: Springer.

Chang, H. 2016. "Pragmatic Realism." *Revista de Humanidades de Valparaíso* 4(2): 107–122.

Chang, H. 2018. "Is Pluralism Compatible with Scientific Realism?" In *The Routledge Handbook of Scientific Realism*, J. Saatsi (ed.), Abingdon: Routledge, 176–186.

Chapman, A. 1989. "Tycho Brahe – Instrument Designer, Observer and Mechanician." *Journal of the British Astronomical Association* 99(2): 70–77.

Christianson, J. 2000. *On Tycho's Island: Tycho Brahe and His Assistants, 1570–1601*. Cambridge: Cambridge University Press.

Churchland, P. M. 1981. "Eliminative Materialism and Propositional Attitudes." *The Journal of Philosophy* 78(2): 67–90.

Churchland, P. M. 1992. "A Deeper Unity: Some Feyerabendian Themes in Neurocomputational Form." In *Connectionism: Theory and Practice*, S. Davis (ed.), Oxford: Oxford University Press, 30–68.

Churchland, P. M. 1997. "To Transform the Phenomena: Feyerabend, Proliferation, and Recurrent Neural Networks." *Philosophy of Science* 64: S408–S420.

Churchland, P. S. 1988. "The Significance of Neuroscience for Philosophy." *Trends in Neurosciences* 11(7): 304–307.

Churchland, P. S. 2002. *Brain-Wise: Studies in Neurophilosophy*. Cambridge: MIT Press.

Clark, S. 2002. "Feyerabend's *Conquest of Abundance*." *Inquiry* 45(2): 249–267.

Clarke, C. 1976. "Space-Time Singularities." *Communications in Mathematical Physics* 49: 17–23.

Clarke, C. 1993. *The Analysis of Space-Time Singularities*. Cambridge: Cambridge University Press.

Cohn, J. 2008. "Citizen Science: Can Volunteers Do Real Research?." *BioScience* 58(3): 192–197.

Collodel, M. 2016. "Was Feyerabend a Popperian? Methodological Issues in the History of the Philosophy of Science." *Studies in History and Philosophy of Science* 57: 27–56.

Copernicus, N. 1543/1995. *On the Revolutions of Heavenly Spheres*. Amherst NY: Prometheus Books.

Couvalis, G. 1987. "Feyerabend's Epistemology and Brecht's Theory of the Drama." *Philosophy and Literature* 11(1): 117–123.

Couvalis, G. 1989. *Feyerabend's Critique of Foundationalism*. Aldershot UK: Avebury.

Crombie, A. 1994. *Styles of Scientific Thinking in the European Tradition: The History of Argument and Explanation Especially in the Mathematical and Biomedical Sciences and Arts*. London: Duckworth.

Curiel, E. 2017. "A Primer on Energy Conditions." In *Towards a Theory of Spacetime Theories*, D. Lehmkuhl, G. Schiemann, and E. Scholz (eds.), Boston: Birkhäuser, 43–104.

Curran, A. 2001. "Brecht's Criticism of Aristotle's Aesthetics of Tragedy." *Journal of Aesthetics and Art Criticism* 59(2): 167–184.

Danto, A. 1986. *Philosophy and the Disenfranchisement of Art*. New York, Columbia University Press.

Danto, A. 1997. *After the End of Art*. Princeton: Princeton University Press.

Daston, L. 2008. "On Scientific Observation." *ISIS*, 99(1): 97–110.

Dawson, G., Lintott, C. and Shuttleworth, S. 2015. "Constructing Scientific Communities: Citizen Science in the Nineteenth and Twenty-First Centuries". *Journal of Victorian Culture* 20(2): 246–254.

Dennett, D. 2006. "Two Steps Closer on Consciousness." In *Paul Churchland*, B. Keeley (ed.), Cambridge: Cambridge University Press, 193–210.

Dennhardt, A., Duerr, A., Brandes, D. and Katzner, T. 2015. "Integrating Citizen-Science Data with Movement Models to Estimate the Size of a Migratory Golden Eagle Population". *Biological Conservation* 184: 68–78.

Dewey, J. 1927. *The Public and Its Problems*. New York: Henry Holt.

Dickinson, J. and R. Bonney. 2012. *Citizen Science: Public Participation in Environmental Research*. Ithaca: Comstock Publishing.

Disch, L. 1996. *Hannah Arendt and the Limits of Philosophy: With a New Preface*. Ithaca: Cornell Press.

Douglas, H. 2000. "Inductive Risk and Values in Science." *Philosophy of Science* 67 (4): 559–79.

Douglas, H. 2005. "Inserting the Public into Science." In *Democratization of Expertise? Exploring Novel Forms of Scientific Advice in Political Decision-Making*, S. Maasen and P. Weingart (eds.), Dordrecht: Springer, 153–169.

Douglas, H. 2009. *Science, Policy, and the Value-Free Ideal*, Pittsburgh: Pittsburgh University Press.

Duhem, P. 1914/1954. *The Aim and Structure of Physical Theory*. Princeton: Princeton University Press.

Dupré, J. 1993. *The Disorder of Things: Metaphysical Foundations for the Disunity of Science*. Harvard: Harvard University Press.

Earman, J. 1995. *Bangs, Crunches, Whimpers, and Shrieks: Singularities and Acausalities in Relativistic Spacetimes*. Oxford: Oxford University Press.

Earman, J., Wüthrich, C., and Manchak, J. 2016. "Time Machines." In *Stanford Encyclopedia of Philosophy*, E. Zalta (ed.),

eBird. 2018. "About eBird". Retrieved online https://ebird.org/about.

Edenhofer, O. and M. Kowarsch. 2015. "Cartography of Pathways: A New Model for Environmental Policy Assessments." *Environmental Science & Policy* 51: 56–64.

Einstein, A. 1934. "On the Method of Theoretical Physics." *Philosophy of Science* 1: 163–169.

Eitzel, M., Cappadonna, J., Santos-Lang, C., Duerr, R., Virapongse, A., West, S., Kyba, C., Bowser, A., Cooper, C., Sforzi, A., Metcalf, A, Harris, E., Theil, M., Haklay, M., Ponciano, L., Ceccaroni, L., Shilling, F., Dorler, D., Heigle, F. Kiessling, T. , Davis, B. and Jiang, Q. 2017. "Citizen Science Terminology Matters: Exploring Key Terms." *Citizen Science: Theory and Practice* 2(1): 1–20.

Elliott, K. 2012. "Epistemic and Methodological Iteration in Scientific Research." *Studies in History and Philosophy of Science* 43: 376–382.

Elliott, K. and J. Rosenberg. 2019. "Philosophical Foundations for Citizen Science." *Citizen Science: Theory and Practice* 4(1): 9.

Ellwood, E., Bart, Jr., H. , Doosey, M., Jue, D., Mann, J., Nelson, G., Rios, N. and Mast, A. 2016. "Mapping Life – Quality Assessment of Novice vs. Expert Georeferencers." *Citizen Science: Theory and Practice* 1(1): 1–12.

Elsner, J. 2006. "From Empirical Evidence to the Big Picture: Some Reflections on Riegl's Concept of *Kunstwollen*." *Critical Inquiry* 32(4): 741–766.

Emmerson, R. 2013. *Key Figures in Medieval Europe, An Encyclopedia*. London: Routledge.

Ens, E., Petina P., Clarke, P., Budden, M., Clubb, L., Doran, B., Douras, C., Gaikwad, J., Gott, B., Leonard, S., Locke, J., Packer, J., Turpin, G. and Wason, S. 2015. "Indigenous Biocultural Knowledge in Ecosystem Science and Management: Review and Insight from Australia." *Biological Conservation* 181: 133–149.

Epstein, S. 1996. *Impure Science: Aids, activism and the politics of knowledge*. Berkeley: University of California Press.

Farrell, R. 2000. "Will the Popperian Feyerabend Please Step Forward: Pluralistic, Popperian Themes in the Philosophy of Paul Feyerabend." *International Studies in the Philosophy of Science* 14(3): 257–266.

Farrell, R. 2003. *Feyerabend and Scientific Values: Tightrope-Walking Rationality*. Dordrecht: Kluwer Academic Publishers.

Fehr, C. and Plaisance, K. 2010. "Socially Relevant Philosophy of Science: An Introduction." *Synthese* 177(3): 301–316.

Feigl, H. 1950. "The Mind-Body Problem in the Development of Logical Empiricism." *Revue Internationale de Philosophie* 11: 64–83.

Feigl, H. 1968/1981. "The *Wiener Kreis* in America." In *Inquiries and Provocations*, R. Cohen (ed.), Dordrecht: Springer, 57–94.

Feyerabend, P. 1948/2016. "The Concept of Intelligibility in Modern Physics (1948)." Translated by Daniel Kuby and Eric Oberheim. *Studies in History and Philosophy of Science Part A* 57: 64–66.

Feyerabend, P. 1951. "Zur Theorie der Basissätze." Dissertation zur Erlangung der Doktorgrades an der philosophischen Fakultät der Universität Wien, Wien: Universität Wien.

Feyerabend, P. 1954a. "Determinismus und Quantenmechanik." *Wiener Zeitschrift für Philosophie, Psychologie, Pädagogik* 5(2): 89–111.

Feyerabend, P. 1954b. "Physik und Ontologie." *Wissenschaft und Weltbild: Monatsschrift für alle Gebiete der Forschung* 7(11/12): 464–480.

Feyerabend, P. 1954a/2015. "Determinism and Quantum Mechanics." In *Physics and Philosophy*, S. Gattei and J. Agassi (eds.), New York: Cambridge University Press, 25–45.

Feyerabend, P. 1954b/2015. "Physics and Ontology." In *Physics and Philosophy*, S. Gattei and J. Agassi (eds.), New York: Cambridge University Press, 9–24.

Feyerabend, P. 1955. "Wittgenstein's Philosophical Investigations." *The Philosophical Review* 64(3), 449–483.

Feyerabend, P. 1956a. "A Note on the Paradox of Analysis." *Philosophical Studies* 7 (6): 92–96.

Feyerabend, P. 1956b. "Eine Bemerkung zum Neumannschen Beweis." *Zeitschrift für Physik* 145(4): 421–423.

Feyerabend, P. 1957a. "On the Quantum-Theory of Measurement." In *Observation and Interpretation: A Symposium of Philosophers and Physicists, Proceedings of the Ninth Symposium of the Colston Research Society*, S. Körner and M. Pryce (eds.), London and New York: Butterworths Scientific Pub, 121–130.

Feyerabend, P. 1957b. "Review of Foundations of Quantum-Mechanics: A Study in Continuity and Symmetry, Alfred Landé." *The British Journal for the Philosophy of Science* 7(28): 354–357.

Feyerabend, P. 1957c. "Zur Quantentheorie der Messung." *Zeitschrift für Physik* 148(5): 551–559.

Feyerabend, P. 1958a. "Mathematical Foundations of Quantum Mechanics, R. T. Beyer (Trans.), Princeton University Press: Princeton (NJ) 1955." *The British Journal for the Philosophy of Science* 8(32): 343–347.

Feyerabend, P. 1958b. "Reichenbach's Interpretation of Quantum-Mechanics." *Philosophical Studies* 9(4): 49–59.

Feyerabend, P. 1958c. "Complementarity." *Proceedings of the Aristotelian Society, Supplementary Volumes* 32: 75–122.

Feyerabend, P. 1958/1981. "An Attempt at a Realistic Interpretation of Experience." In P. Feyerabend, *Philosophical Papers Volume 1: Realism, Rationalism and Scientific Method*, Cambridge: Cambridge University Press, 17–36.

Feyerabend, P. 1960a. "On the Interpretation of Scientific Theories." In *Proceedings of the 12th International Congress of Philosophy, Venice,*

12–18 September 1958, Logic, Theory of Knowledge, Philosophy of Science, Philosophy of Language 5. Florence: Sansoni, 151–159.

Feyerabend, P. 1960b. "Professor Landé on the Reduction of the Wave Packet." *American Journal of Physics* 28(5): 507.

Feyerabend, P. 1960c. "Professor Bohm's Philosophy of Nature." *The British Journal for the Philosophy of Science* 10(40): 321–338.

Feyerabend, P. 1960/1981. "On the Interpretation of Scientific Theories." In *Philosophical Papers Volume 1: Realism, Rationalism and Scientific Method*, P. Feyerabend (ed.). Cambridge: Cambridge University Press, 37–43.

Feyerabend, P. 1961. "Niels Bohr's Interpretation of the Quantum Theory." In *Current Issues in the Philosophy of Science: Symposia of Scientists and Philosophers*, H. Feigl and G. Maxwell (eds.), Proceedings of Section L of the American Association for the Advancement of Science. New York: Holt, Rinehart and Winston, 371–390.

Feyerabend, P. 1961/1981. "Knowledge Without Foundations." In *Realism, Rationalism and Scientific Method*, P. Feyerabend (ed.), Cambridge: Cambridge University Press, 50–78.

Feyerabend, P. 1962a. "Explanation, Reduction, and Empiricism." In *Scientific Explanation, Space, and Time*, H. Feigl and G. Maxwell (eds.), 3. Minnesota Studies in the Philosophy of Science. Minneapolis: University of Minnesota Press, 28–97.

Feyerabend, P. 1962b. "Problems of Microphysics." In *Frontiers of Science and Philosophy*, R. Colodny (ed.), University of Pittsburgh Series in the Philosophy of Science 1. Pittsburgh: University of Pittsburgh Press, 189–283.

Feyerabend, P. 1963a. "Materialism and the Mind-Body Problem." *The Review of Metaphysics*, 17(1): 49–66.

Feyerabend, P. 1963b. "Über konservative Züge in den Wissenschaften und insbesondere in der Quantentheorie und ihre Beseitigung." *Club Voltaire. Jahrbuch Für Kritische Aufklärung* 1: 280–93.

Feyerabend, P. 1964. "Realism and Instrumentalism: Comments of the Logic of Factual Support." In *The Critical Approach to Science and Philosophy: In Honour of Karl R. Popper*, Mario Bunge (ed.), London and New York: The Free Press of Glencoe, 280–308.

Feyerabend, P. 1964/1981. "Realism and Instrumentalism: Comments on the Logic of Factual Support." In *Philosophical Papers Volume 1: Realism, Rationalism and Scientific Method*, P. Feyerabend (ed.), Cambridge: Cambridge University Press, 176–202.

Feyerabend, P. 1965a. "Problems of Empiricism." In *Beyond the Edge of Certainty: Essays in Contemporary Science and Philosophy*, R. Colodny (ed.), 2. University of Pittsburgh Series in the Philosophy of Science. Englewood Cliffs (NJ): Prentice-Hall, 145–260.

Feyerabend, P. 1965b. "Reply to Criticism: Comments on Smart, Sellars and Putnam." In *Boston Studies in the Philosophy of Science, Vol 2*. New York: Humanities Press, 223–261.

Feyerabend, P. 1965/1981. "Reply to Criticism: Comments on Smart, Sellars, and Putnam." In *Realism, Rationalism, and Scientific Method: Philosophical Papers, Volume 1*, P. Feyerabend (ed.), Cambridge: Cambridge University Press, 104–131.

Feyerabend, P. 1966. "The Structure of Science." *The British Journal for the Philosophy of Science* 17(3): 237–249.

Feyerabend, P. 1967a. "On the Improvement of the Sciences and the Arts, and the Possible Identity of the Two." In *Boston Studies in the Philosophy of Science*, vol. 3, R. Cohen and M. Wartofsky (eds.), Dordrecht: Reidel, 387–415.

Feyerabend, P. 1967b. "The Theatre as an Instrument of the Criticism of Ideologies: Notes on Ionesco." *Inquiry* 10: 298–312.

Feyerabend, P. 1968a. "On a Recent Critique of Complementarity: Part I." *Philosophy of Science* 35(4): 309–331.

Feyerabend, P. 1968b. "Science, Freedom, and the Good Life." *The Philosophical Forum* 1(2): 127–135.

Feyerabend, P. 1968/1999. "Outline of a Pluralistic Theory of Knowledge and Action." In *Knowledge, Science and Relativism. Philosophical Papers Volume 3*, J. Preston (ed.), Cambridge: Cambridge University Press, 104–111.

Feyerabend, P. 1969a. "Linguistic Arguments and Scientific Method." *Telos* 3: 43–63.

Feyerabend, P. 1969b. "On a Recent Critique of Complementarity: Part II." *Philosophy of Science* 36(1): 82–105.

Feyerabend, P. 1969c. "Science without Experience." *The Journal of Philosophy* 66 (22): 791–794.

Feyerabend, P. 1970a. "Consolations for the Specialist." In *Criticism and the Growth of Knowledge*, I. Lakatos and A. Musgrave (eds.), Cambridge: Cambridge University Press, 197–230.

Feyerabend, P. 1970b. "Experts in a Free Society." *The Critic* 29(2): 58–69.

Feyerabend, P. 1970c. "Problems of Empiricism, Part II." In *The Nature and Function of Scientific Theories: Essays in Contemporary* Science *and Philosophy*, University of Pittsburgh Series in the Philosophy of Science, Vol. 4, R. Colodny (ed.), Pittsburgh: University of Pittsburgh Press, 275–353.

Feyerabend, P. 1970d. "Classical Empiricism." In *The Methodological Heritage of Newton*, R. Butts (ed.), Oxford: Blackwell Press, 150–166.

Feyerabend, P. 1970e. "Philosophy of Science: A Subject with a Great Past." *Historical and Philosophical Perspectives of Science* 5: 173–182.

Feyerabend, P. 1972. "On the Limited Validity of Methodological Rules." In *Philosophical Papers. Vol. 3, Knowledge, Science and Relativism*, J. Preston (ed.), Cambridge: Cambridge University Press, 138–180.

Feyerabend, P. 1973/1999. "Theses on Anarchism." In *For and Against Method*, M. Motterlini (ed.), Chicago: University of Chicago Press, 113–118.

Feyerabend, P. 1975a. *Against Method: Outline of an Anarchistic Theory of Knowledge* (1st Edition). London: New Left Books.

Feyerabend, P. 1975b. "Let's Make More Movies." In *The Owl of Minerva: Philosophers on Philosophy*, J. Bontempo and S. Odell (eds.), New York: McGraw-Hill, 201–210.

Feyerabend, P. 1975c. "How to Defend Society Against Science." *Radical Philosophy* 11: 3–8.

Feyerabend, P. 1975d. "'Science.' The Myth and its Role in Society." *Inquiry: An Interdisciplinary Journal of Philosophy* 18(2): 167–181.

Feyerabend, P. 1975/1988. *Against Method: Outline of an Anarchistic Theory of Knowledge* (2nd Edition). London: Verso Books.

Feyerabend, P. 1975/1993. *Against Method: Outline of an Anarchistic Theory of Knowledge* (3rd Edition). London: Verso Books.

Feyerabend, P. 1975/2010. *Against Method: Outline of an Anarchistic Theory of Knowledge* (4th Edition). London: Verso.

Feyerabend, P. 1976. "On the Critique of Scientific Reason." In *Method and Appraisal in the Physical Sciences: The Critical Background to Modern* Science, *1800–1905*, C. Howson (ed.), Cambridge: Cambridge University Press, 309–339.

Feyerabend, P. 1977. "Changing Patterns of Reconstruction." *The British Journal for the Philosophy of Science* 28: 351–369.

Feyerabend, P. 1978a. *Science in a Free Society*. London: New Left Books.

Feyerabend, P. 1978b. "From Incompetent Professionalism to Professionalized Incompetence–The Rise of a New Breed of Intellectuals." *Philosophy of the Social Sciences* 8(1): 37–53.

Feyerabend, P. 1981a. "Introduction to Volumes 1 and 2." In *Philosophical Papers Volume 1: Realism, Rationalism and Scientific Method*, P. Feyerabend (ed.), Cambridge: Cambridge University Press, ix–xiv.

Feyerabend, P. 1981b. "Introduction: Scientific Realism and Philosophical Realism." In *Philosophical Papers Volume 1: Realism, Rationalism and Scientific Method*, P. Feyerabend (ed.), Cambridge: Cambridge University Press, 3–16.

Feyerabend, P. 1981c. "Proliferation and Realism as Methodological Principles." In *Rationalism, Realism, and Scientific Method: Philosophical Papers, Vol. 1*, P. Feyerabend (ed.), Cambridge: Cambridge University Press, 139–145.

Feyerabend, P. 1981d. "Review: More Clothes from the Emperor's Bargain Basement." *The British Journal for the Philosophy of Science* 32(1): 57–71.

Feyerabend, P. 1981e. "Historical Background: Some Observations on the Decay of the Philosophy of Science." In *Problems of Empiricism: Philosophical Papers, Volume 2*, P. Feyerabend, (ed.), Cambridge: Cambridge University Press, 1–33.

Feyerabend, P. 1984. *Scienza Come Arte*. Transl. Libero Sosio. Bari: Laterza.

Feyerabend, P. 1987a. *Farewell to Reason*. London and New York: Verso.

Feyerabend, P. 1987b. "Creativity: A Dangerous Myth." *Critical Inquiry* 13(4): 700–711.

Feyerabend, P. 1988. "Knowledge and the Role of Theories". *Philosophy of the Social Science* 18: 157–178.

Feyerabend, P. 1989/1999. "Realism and the Historicity of Knowledge." In *Conquest of Abundance: A Tale of Abstraction versus the Richness of Being*,

P. Feyerabend and B. Terpstra (ed.), Chicago: University of Chicago Press, 131–146.

Feyerabend, P. 1991. "Concluding Unphilosophical Conversation." In *Beyond Reason: Essays on the Philosophy of Paul Feyerabend*, G. Munévar (ed.), London: Kluwer, 433–448.

Feyerabend, P. 1992. "Nature as a Work of Art." *Common Knowledge* 1(3): 3–9.

Feyerabend, P. 1992/1999. "Historical Comments on Realism." In *Conquest of Abundance: A Tale of Abstraction versus the Richness of Being*, P. Feyerabend and B. Terpstra (eds.), Chicago: University of Chicago Press, 197–205.

Feyerabend, P. 1993. "Not a Philosopher." In *Falling in Love with Wisdom: American Philosophers Talk about Their Calling*, D. Karnos and R. Shoemaker (eds.), Oxford: Oxford University Press, 16–17.

Feyerabend, P. 1994a. "Concerning an Appeal for Philosophy." *Common Knowledge* 3: 10–13.

Feyerabend, P. 1994b. "Art as a Product of Nature as a Work of Art." *World Futures: Journal of General Evolution* 40(1–3): 87–100.

Feyerabend, P. 1994/1999. "Realism." In *Conquest of Abundance: A Tale of Abstraction versus the Richness of Being*, P. Feyerabend and B. Terpstra (eds.), Chicago: University of Chicago Press, 178–196.

Feyerabend, P. 1995. *Killing Time: The Autobiography of Paul Feyerabend*. Chicago: University of Chicago Press.

Feyerabend, P. 1995/1999. "What Reality?" In *Conquest of Abundance: A Tale of Abstraction versus the Richness of Being*, P. Feyerabend and B. Terpstra (eds.), Chicago: University of Chicago Press, 206–216.

Feyerabend, P. 1996. "Theoreticians, Artists Artisans." *Leonardo* 29(1): 23–28.

Feyerabend, P. 1999. *Conquest of Abundance: A Tale of Abstraction versus the Richness of Being*. Chicago: University of Chicago Press.

Feyerabend, P. 2011. *The Tyranny of Science*. Cambridge: Polity Press.

Feyerabend, P. 2015. *Philosophical Papers. Vol. 4, Physics and Philosophy*. Cambridge: Cambridge University Press.

Feyerabend, P. 2016. *Philosophy of Nature*. E. Oberheim and H. Heit (eds.), Cambridge: Polity.

Feyerabend, P. 2020. Feyerabend's Formative Years. Volume 1. Feyerabend and Popper: Correspondences and Unpublished Papers. Matteo Collodel and Eric Oberheim (eds.), Vienna Circle Institute Library. Vienna-Berlin-Münster: Springer.

Field, J. 2004. *Piero Della Francesca, A Mathematician's Art*. Cambridge: Cambridge University Press.

Field, J. 2016. "The Unhelpful Notion of Renaissance Man." *Interdisciplinary Science Reviews* 25(2): 188–201.

Finocchiaro, M. 2001. "Science, Religion, and the Historiography of the Galileo Affair: On the Undesirability of Oversimplification." *Osiris* 16: 114–132.

Fleck, L. 1935/1981. *Genesis and Development of a Scientific Fact*. Chicago: Chicago: University of Chicago Press.

Franklin, A. 1993. *The Rise and Fall of the Fifth Force: Discovery, Pursuit, and Justification in Modern Physics*. New York: American Institute of Physics.

Freitag, A., Meyer, R. and Whiteman, L. 2016. "Strategies Employed by Citizen Science Programs to Increase the Credibility of Their Data." *Citizen Science: Theory and Practice* 1(1): 1–11.

Frensley, T., Crall, A., Stern, M., Jordan, R., Gray, S., Prysby, M., Newman, G., Hmelo-Silver, C., Mellor, D. and Huang, J. 2017. "Bridging the Benefits of Online and Community Supported Citizen Science: A Case Study on Motivation and Retention with Conservation-Oriented Volunteers." *Citizen Science: Theory and Practice* 2(1): 1–14.

Gade, J. A. 1947. *The Life and Times of Tycho Brahe*. Princeton: Princeton University Press.

Galaxy Zoo: Clump Scout. 2019. Zooniverse. Accessed Online on August 10, 2019. www.zooniverse.org/projects/hughdickinson/galaxy-zoo-clump-scout

Galilei, G. 1632/2001. *Dialogue Concerning the Two Chief World Systems: Ptolemaic and Copernican*. New York: The Modern Library.

Galison, P. and Stump, D. J. (eds.) 1996. *The Disunity of Science: Boundaries, Context, and Power*. Stanford: Stanford University Press.

Galloway, G. and Ling, E. 2017. "Some Remarks on the C^0-(In)Extendibility of Spacetimes." *Annales Henri Poincaré* 18: 3427–3447.

Geelan, D. 2001. "Feyerabend Revisited: Epistemological Anarchy and Disciplines Eclecticism in Educational Research." *Australian Educational Researcher* 28(1): 129–146.

Genet, K. and Sargent, L. 2003. "Evaluation of Methods and Data Quality from a Volunteer-Based Amphibian Call Survey." *Wildlife Society Bulletin* 31(3): 703–714.

Geroch, R. 1970. "Singularities." In *Relativity* M. Carmeli, S. Fickler and L. Witten (eds.), New York: Plenum Press, 259–291.

Geroch, R. 1977. "Prediction in General Relativity." In *Foundations of Space-Time Theories, Minnesota Studies in the Philosophy of Science Vol. VIII*, J. Earman, C. Glymour and J. Stachel (eds.), Minneapolis: University of Minnesota Press, 81–93.

Geroch, R. and Horowitz, G. 1979. "Global Structure of Spacetimes." In *General Relativity: An Einstein Centenary Survey*, S. Hawking and W. Israel (eds.), Cambridge: Cambridge University Press, 212–293.

Giere, R. 2006. *Scientific Perspectivism*. Chicago: University of Chicago Press.

Gillies, D. 2019. "Lakatos, Popper, and Feyerabend: Some Personal Reminiscences." *Dilemata* 29: 93–108.

Gingerich, O. and J. Voelkel. 1998. "Tycho Brahe's Copernican Campaign." *Journal for the History of Astronomy* 29 (1): 1–34.

Ginzburg, C. 1998. "Style as Inclusion, Style as Exclusion." In *Picturing Art, Producing Science*, P. Galison and C. Jones (eds.), New York: Routledge, 27–54.

Glymour, C. 1977. "Indistinguishable Space-Times and the Fundamental Group." In *Foundations of Space-Time Theories, Minnesota Studies in the Philosophy of*

Science Vol. VIII, J. Earman, C. Glymour and J. Stachel (eds.), Minneapolis: University of Minnesota Press, 50–60.

Godfrey-Smith, P. 2003. *Theory and Reality*. Chicago: University of Chicago Press.

Golumbic, Y., Orr, D., Baram-Tsabari, A. and Fishbain, B. 2017. "Between Vision and Reality: A Study of Scientists' Views on Citizen Science." *Citizen Science: Theory and Practice* 2(1): 1–13.

Gombrich, E. 1960/2002. *Art and Illusion*. London: Phaidon.

Gombrich, E. 1972. "The 'What' and the 'How': Perspective Representation and the Phenomenal World." In *Logic and Art: Essays in Honour of Nelson Goodman*, R Rudner and I. Scheffler (eds.), Indianapolis: Bobbs Merrill, 129–149.

Gombrich, E. 1975. *The Sense of Order*. London: Phaidon.

Gombrich, E. H. 1993. *A Lifelong Interest*. London: Thames and Hudson.

Goodman, N. 1976. *Languages of Art*. Indianapolis: Hackett.

Gutting, G. (ed.) 2005. *Continental Philosophy of Science*, Oxford: Blackwell.

Haack, S. 2007. *Defending Science – Within Reason: Between Scientism and Cynicism*. Amherst: Prometheus Books.

Hacking, I. 1982. "Language, Truth and Reason." In *Rationality and Relativism*, M. Hollis and S. Lukes (eds.), Cambridge: MIT Press, 48–66.

Hacking, I. 1983. *Representing and Intervening*. Cambridge: Cambridge University Press.

Hacking, I. 1991. "Speculation, Calculation and the Creation of Phenomena." In *Beyond Reason: Essays on the Philosophy of Paul K. Feyerabend*, G. Munévar (ed.), London: Kluwer, 131–158.

Hacking, I. 1992. "Style for the Historian and the Philosopher." *Studies in History and Philosophy of Science* 23(1): 1–20.

Hacking, I. 1996. "The Disunities of the Sciences." In *The Disunity of Science*, P. Galison and D. Stump (eds.), Stanford: Stanford University Press, 37–74.

Hacking, I. 2012. "'Language, Truth and Reason' Thirty Years Later." *Studies in History and Philosophy of Science* 43(4): 599–609.

Halliwell, S. 1998. *Aristotle's Poetics*. Chicago: University of Chicago Press.

Halliwell, S. 2002. *The Aesthetics of Mimesis*. Princeton: Princeton University Press.

Havstad, J. and M. J. Brown. 2017. "Inductive Risk, Deferred Decisions, and Climate Science Advising." In *Exploring Inductive Risk*, K. Elliott and T. Richards (eds.), Oxford: Oxford University Press, 101–123.

Hawking, S. 1969. "The Existence of Cosmic Time Functions." *Proceedings of the Royal Society A* 308: 433–435.

Hawking, S. and Ellis, G. 1973. *The Large Scale Structure of Space-Time*. Cambridge: Cambridge University Press.

Heidegger, M. 1971. *Poetry, Language, Thought*, A. Hofstadter (ed.), New York: Harper & Row.

Heidegger, M. 1977. *The Question Concerning Technology (and other Essays)*. W. Lovitt (trans.), New York: Harper & Row.

Heilbron, J. 1992. "Creativity and Big Science." *Physics Today* 45: 42–47.

Heller, L. 2016. "Between Relativism and Pluralism: Philosophical and Political Relativism in Feyerabend's Late Work." *Studies in History and Philosophy of Science* 57: 96–105.

Hemingway, A. 2009. "E.H. Gombrich in 1968: Methodological Individualism and the Contradictions of Conservatism." *Human Affairs: A Postdisciplinary Journal for Humanities & Social Sciences* 19(3): 297–303.

Hempel, C. 1950. "Problems and Changes in the Empiricist Criterion of Meaning." *Révue Internationale de Philosophie* 4, 41–63.

Hempel, C., 1952. "Fundamentals of Concept Formation in Empirical Science." In *Foundations of the Unity of Science*, Volume 2, O. Neurath, R. Carnap, and C. Morris (eds.), Chicago: University of Chicago Press, 651–746.

Henderson, H. 1974. "Information and the New Movements for Citizen Participation." *Annals of the American Academy of Political and Social Science* 412: 34–43.

Henderson, S. 2012. "Citizen Science Comes of Age." *Frontiers in Ecology and the Environment* 10(6): 283.

Hertz, H. 1894/1899. *The Principles of Mechanics*, Presented in a New Form, D. Jones and J. Walley (trans.), London: Macmillan.

Hirsh, E. 2016. "Three Degrees of Carnapian Tolerance." In *Ontology After Carnap*, S. Blatti and S. Lapointe (eds.), Oxford: Oxford University Press, 105–122.

Howard, D. 2004. "Who Invented the 'Copenhagen Interpretation'? A Study in Mythology." *Philosophy of Science* 71(5): 669–682.

Hoyningen-Huene, P. 1987. "Context of Discovery and Context of Justification." *Studies in History and Philosophy of Science* 18(4): 501–515.

Hoyningen-Huene, P. 2002. "Paul Feyerabend und Thomas Kuhn." *Journal for General Philosophy of Science* 33(1): 61–83.

Hoyningen-Huene, Paul. 2006. "More Letters by Paul Feyerabend to Thomas S. Kuhn on Proto-Structure." *Studies in History and Philosophy of Science* 37(4): 610–32.

Husserl, E. 1970. *The Crisis of European Sciences and Transcendental Phenomenology*, D. Carr (trans.), Evanston: Northwestern University Press.

Invasive Plant Atlas of New England. 2018. "Distribution Maps". Retrieved Online www.eddmaps.org/about. Last date accessed: November 18, 2020.

Ioannidis, J. 2005. "Why Most Published Research Findings Are False." *PLoS Medicine* 2(8): e124.

Iversen, M. 1993. *Alois Riegl: Art History and Theory*. Cambridge: Cambridge University Press.

Jackson, M., Gergel, S. and Martin, K. 2015. "Citizen Science and Field Survey Observations Provide Comparable Results for Mapping Vancouver Island White-tailed Ptarmigan (Lagopus saxatilis) Distributions". *Biological Conservation* 181: 162–172.

Jacobs, S. 2003. "Misunderstanding John Stuart Mill on Science: Paul Feyerabend's Bad Influence." *The Social Science Journal* 40(2): 201–212.

James, W. 1956. "The Will to Believe." In *The Will to Believe and Human Immortality*, W. James (ed.), New York: Dover Publications, 1–31.

James, W. 1907/1978. *Pragmatism* and *The Meaning of Truth*. Cambridge: Harvard University Press.

Jeffrey, R. 1956. "Valuation and Acceptance of Scientific Hypotheses." *Philosophy of Science* 23(3): 237–246.

Jordan, R., Gray, S., Howe, D. and Brooks, W. 2011. "Knowledge Gain and Behavioural Change in Citizen-Science Programs." *Conservation Biology* 25(6): 1148–1154.

Keegan, J. 1976. *The Face of Battle*. London: Jonathan Cape.

Keeley, B. 2006. "Introduction: Becoming Paul M. Churchland (1942–)." In *Paul Churchland*, B. Keeley (ed.), Cambridge: Cambridge University Press, 1–32.

Kellert, S., Longino, H and Waters, K. 2006. *Scientific Pluralism*. Minneapolis: University of Minnesota Press.

Kelty, C. and A. Panofsky. 2014. "Disentangling Public Participation in Science and Biomedicine." *Genome Medicine* 6(8): 1–14.

Kepler, J. 1609/2004. *Selections from Kepler's Astronomia Nova: A Science Classics Module for Humanities Studies*, selected, translated, and annotated by W. Donahue. Santa Fe: Green Lion Press.

Kidd, I.J. (unpublished manuscript). "Reassessing Epistemological Dadaism: Feyerabend on Theory and Practice in Art and Science". Available at: www .academia.edu/1665818/Reassessing_Epistemological_Dadaism_Feyerabend_o n_Theory_and_Practice_in_Art_and_Science, last accessed August 16, 2018.

Kidd, I. J. 2010. *Pluralism and the 'Problem of Reality' in the Later Philosophy of Paul Feyerabend*, Durham thesis, Durham University.

Kidd, I.J. 2011a. "Objectivity, Abstraction, and the Individual: The Influence of Søren Kierkegaard on Paul Feyerabend." *Studies in History and Philosophy of Science* 42(1): 125–134.

Kidd, I. J. 2011b. "Rethinking Feyerabend: The 'Worst Enemy of Science'?." *PLoS Biology* 9(10).

Kidd, I.J. 2012. "Feyerabend, Pseudo-Dionysius, and the Ineffability of Reality." *Philosophia* 40: 365–377.

Kidd, I.J. 2013a. "Feyerabend on Science and Education." *Journal of Philosophy of Education* 47(3): 407–422.

Kidd, I.J. 2013b. "A Pluralist Challenge to 'Integrative Medicine': Feyerabend and Popper on the Cognitive Value of Alternative Medicine." *Studies in History and Philosophy of Biological and Biomedical Science* 44(3): 392–400.

Kidd, I.J. 2016a. "Why Did Feyerabend Defend Astrology? Integrity, Virtue, and the Authority of Science." *Social Epistemology* 30(4): 464–482.

Kidd, I.J. 2016b. "Was Feyerabend a Postmodernist?" *International Studies in Philosophy of Science* 30(1): 1–14.

Kidd, I.J. 2016c. "'What's So Great about Science?' Feyerabend on Science, Ideology, and the Cold War." In *Science Studies During the Cold War and Beyond*, E. Aronova and S. Turchetti (eds.), Basingstoke: Palgrave-Macmillan, 55–76.

Kidd, I.J. 2016d. "Inevitability, Contingency, and Epistemic Humility." *Studies in History and Philosophy of Science* 55: 12–19.

Kidd, I.J. 2017a. "Reawakening to Wonder: Wittgenstein and Feyerabend on Scientism." In *Wittgenstein on Scientism*, J. Beale and I.J Kidd (eds.), London: Routledge, 101–115.

Kidd, I.J. 2017b. "Other Histories, Other Sciences." *Studies in History and Philosophy of Science* 61: 57–60.

Kidd, I.J. 2018. "Feyerabend, Pluralism, and Parapsychology." *Bulletin of the Parapsychology Association* 10(1): 5–9.

Kim, J. 1998. "The Mind–Body Problem After Fifty Years." *Royal Institute of Philosophy Supplements* 43: 3–21.

Kitcher, P. 1993. *The Advancement of Science: Science Without Legend, Objectivity Without Illusions*. Oxford: Oxford University Press.

Kitcher, P. 2019. "So . . . Who is Your Audience?" *European Journal for Philosophy of Science* 9(1): 1–15.

Kolok, A., Schoenfuss, H., Propper, C. and Vail, T. 2011. "Empowering Citizen Scientists: The Strength of Many in Monitoring Biologically Active Environmental Contaminants." *BioScience* 61(8): 626–630.

Körner, S (ed). 1957. *Observation and Interpretation: A Symposium of Philosophers and Physicists, Proceedings of the Ninth Symposium of the Colston Research Society*. New York: Butterworths Scientific Pub.

Kožnjak, B. 2017. "The Missing History of Bohm's Hidden Variables Theory: The Ninth Symposium of the Colston Research Society, Bristol, 1957." *Studies in History and Philosophy of Science Part B: Studies in History and Philosophy of Modern Physics* 62: 85–97.

Kresge, S. 1996. "Feyerabend Unbound." *Philosophy of the Social Sciences* 26(2): 293–303.

Krimsky, S. 2003. *Science in the Private Interest: Has the Lure of Profits Corrupted Biomedical Research?* Lanham: Rowman & Littlefield.

Kripke, S. 1972. "Naming and Necessity." In *Semantics of Natural Language*, D. Davidson and G. Harman (eds.), Dordrecht: Springer, 253–355.

Kuby, D. 2010. "Paul Feyerabend in Wien 1946–1955: Das Österreichische College und der Kraft-Kreis." In *Auf der Suche nach authentischem Philosophieren. Philosophie in Österreich 1951–2000. Verdrängter Humanismus-verzögerte Aufklärung*, M. Benedikt, R. Knoll, F. Schwediauer, and C. Zehetner (eds.), 1041–1056. Verdrängter Humanismus – verzögerte Aufklärung: Philosophie in Österreich von 1400 bis heute 6. Wien: WUV.

Kuby, D. 2015, "Feyerabend, Paul (1924–94)", *International Encyclopedia of Social and Behavioural Sciences* (2nd Edition). Amsterdam: Elsevier, 117–123.

Kuby, D. 2016. "Feyerabend's 'The Concept of Intelligibility in Modern Physics' (1948)." *Studies in History and Philosophy of Science Part A* 57: 57–63.

Kuby, D. 2018. "Carnap, Feyerabend, and the Pragmatic Theory of Observation." *HOPOS* 8: 432–470.

Kuby, D. 2019a. "Against the Historical Turn in Philosophy of Science: The Case of Feyerabend's Early Critique of Kuhn." (In preparation).

Kuby, D. (2020). "Decision-Based Epistemology: sketching a systematic framework of Feyerabend's metaphilosophy." *Synthese*, 1–29.

Kuhn, T. 1977. "Objectivity, Value Judgment, and Theory Choice." In *The Essential Tension*, T. Kuhn (ed.), Chicago: University of Chicago Press, 320–339.

Kuhn, T. 1962/2012. *The Structure of Scientific Revolutions* (4th Edition). Chicago: University of Chicago Press.

Kuhn, T. 1963. "The Function of Dogma in Scientific Research." In *Scientific Change*, A. Crombie (ed.), London: Heinemann, 347–69.

Kulka, T. 1977. "How Far Does Anything Go?: Comments on Feyerabend's Epistemological Anarchism." *Philosophy of the Social Sciences* 7: 277–287.

Kusch, M. 2016. "Relativism in Feyerabend's Later Writings." *Studies in History and Philosophy of Science* A57: 106–113.

Kwa, C. 2012. "An 'Ecological' View of Styles of Science and Of Art: Alois Riegl's Explorations of the Style Concept." *Studies in History and Philosophy of Science Part A* 43: 610–618.

Lakatos, I. 1970. "Falsification and the Methodology of Scientific Research Programmes." In *Philosophical Papers: Volume 1. The Methodology of Scientific Research Programmes*, J. Worrall and G. Currie (eds.), Cambridge: Cambridge University Press, 8–101.

Lakatos, I. 1971. "History of Science and Its Rational Reconstructions." In *Philosophical Papers: Volume 1. The Methodology of Scientific Research Programmes*, J. Worrall and G. Currie (eds.), Cambridge: Cambridge University Press, 102–138.

Lakatos, Imre and Zahar, Elie. 1976. Why Did Copernicus's Research Programme Supersede Ptolemy's? In Lakatos, Imre. (1978) *Philosophical Papers: Volume 1. The Methodology of Scientific Research Programmes*. Cambridge University Press, 168–192.

Latour, B. 1987. *Science in Action: How to Follow Scientists and Engineers Through Society*. Cambridge: Harvard University Press.

Latour, B. 2004. *Politics of Nature: How to Bring the Sciences into Democracy*. Cambridge MA: Harvard University Press.

Laudan, L. 1977. *Progress and Its Problems*. Berkeley: University of California Press.

Laudan, L. 1984. *Science and Values: The Aims of Science and Their Role in Scientific Debate*. Berkeley: University of California Press.

Laudan, L. 1989. "The Rational Weight of the Scientific Past: Forging Fundamental Change in a Conservative Discipline." In *What the Philosophy of Biology Is*, M. Ruse (ed.), Nijhoff International Philosophy Series 32. Springer Netherlands, 209–20.

Laudan, L. 1990. "Normative Naturalism." *Philosophy of Science* 57(1): 44–59.

Link, W. and J. Sauer. 2007. "Seasonal Components of Avian Population Change: Joint Analysis of Two Large-Scale Monitoring Programs". *Ecology* 88: 49–55.

Lipton, P. 2004. "Epistemic Options." *Philosophical Studies* 121: 147–158.

Lloyd, E. 1996. "The Anachronistic Anarchist." *Philosophical Studies* 81(2–3): 247–261.

Lloyd, E. 1997. "Feyerabend, Mill, and Pluralism." *Philosophy of Science* 64: S396–S407.

Louisiana Bucket Brigade Press Release. 2000. "Troubled Orion Refinery Violates State Air Standards: Citizen Air Sample Exposes Release of Carcinogen." Retrieved online at www.labucketbrigade.org. Last date accessed: November 18, 2020.

Low, R. 2012. "Time Machines, Maximal Extensions and Zorn's Lemma." *Classical and Quantum Gravity* 29: 097001.

Lugg, A. 1984. "Review: Changing Fortunes of the Method of Hypothesis." *Erkenntnis* 21: 433–438.

Mach, E. 1911. *History and Root of the Principle of the Conservation of Energy*, P. Jourdain (trans.). Chicago: Open Court Publishing.

Magee, P. 2018. *Faulty Instruments, the Sea and the Stars: A Digital Fine Art Practice*. Ph.D. thesis, University College London.

Maia Neto, J. 1991. "Feyerabend's Scepticism." *Studies in the History and Philosophy of Science* 22(4): 543–555.

Malament, D. 1977. "Observationally Indistinguishable Space-Times." In *Foundations of Space-Time Theories, Minnesota Studies in the Philosophy of Science Vol. VIII*, J. Earman, C. Glymour, and J. Stachel (eds.), Minneapolis: University of Minnesota Press, 61–80.

Malament, D. 2012. *Topics in the Foundations of General Relativity and Newtonian Gravitation Theory*. Chicago: University of Chicago Press.

Manchak, J. 2009. "Can We Know the Global Structure of Spacetime?" *Studies in History and Philosophy of Modern Physics* 40: 53–56.

Manchak, J. 2011. "What is a Physically Reasonable Spacetime?" *Philosophy of Science* 78: 410–420.

Manchak, J. 2013. "Global Spacetime Structure." In *The Oxford Handbook of Philosophy of Physics*, R. Batterman (ed.), Oxford: Oxford University Press, 587–606.

Manchak, J. 2016. "Is the Universe As Large as It Can Be?" *Erkenntnis* 81: 1341–1344.

Manchak, J. 2017. "On the Inextendibility of Space-Time." *Philosophy of Science* 84: 1215–1225.

Martin, E. 2016. "Late Feyerabend on Materialism, Mysticism, and Religion." *Studies in History and Philosophy of Science Part A* 57: 129–136.

Martin, R. 1991. *Pierre Duhem: Philosophy and History in the Work of a Believing Physicist*. La Salle: Open Court.

McMullin E. 2007. "Taking an Empirical Stance," In *Images of Empiricism: Essays on Science and Stances, with a Reply from Bas C. van Fraassen*, B. Monton (ed.), Oxford, New York: Oxford University Press, 167–182.

Mendelovici, A. 2018. *The Phenomenal Basis of Intentionality*. New York: Oxford University Press.

Meyer, A. 2011. "On the Nature of Scientific Progress: Anarchistic Theory Says 'Anything Goes'–But I Don't Think So." *PLoS Biology* 9(10): 1–4.

Meynell, H. 1978. "Feyerabend's Method." *The Philosophical Quarterly* 28(112): 242–252.

Mill, J. 1859/1956. *On Liberty*. New York: The Liberal Arts Press.

Miller-Rushing, A, Primack, R. and Bonney, R. 2012. "The History of Public Participation in Ecological Research." *Ecology and Environment* 10:285–290.

Misak, C. 2018. *Cambridge Pragmatism: From Peirce and James to Ramsey and Wittgenstein*. Oxford: Oxford University Press.

Mitchell, S. 2004. "The Prescribed and Proscribed Values in Science Policy." In *Science, Values, and Objectivity*, P. Machamer and G. Wolters (eds.), Pittsburgh: University of Pittsburgh Press, 245–255.

Mosley, A. 2007. *Bearing the Heavens: Tycho Brahe and the Astronomical Community of the Late Sixteenth Century*. Cambridge: Cambridge University Press.

Motion, A. 2019. "What Can Citizen Science Do For Us?" Chemistry World. Accessed online October 12, 2019. www.chemistryworld.com/opinion/what-can-citizen-science-do-for-us/3010269.article.

Motterlini, M. (ed.). 1999. *For and Against Method, Including Lakatos's Lectures on Scientific Method, and the Lakatos-Feyerabend Correspondence*. Chicago: University of Chicago Press.

Motterlini, M. 2002. "Reconstructing Lakatos: A Reassessment of Lakatos' Epistemological Project in the Light of the Lakatos Archive." *Studies in the History and Philosophy of Science* 33: 487–509.

Mount, H. 2014. "Gombrich and the Fathers of Art History." In *Meditations on a Heritage*, P. Taylor (ed.), London: The Warburg Institute and Paul Holberton Publishing, 22–35.

Mueller, M., Tippins, D. and Bryan, L. 2012. "The Future of Citizen Science." *Democracy & Education* 20(1): 1–12.

Muenich, R., Peel, S., Bowling, L., Haas, M., Turco, R., Frankenberger, J. and Chaubey, I. 2016. "The Wabash Sampling Blitz: A Study on the Effectiveness of Citizen Science." *Citizen Science: Theory and Practice* 1(1): 1–15.

Munévar, G. 1991. "Science in Feyerabend's Free Society." In *Beyond Reason: Essays on the Philosophy of Paul Feyerabend*, G. Munévar (ed.), Boston: Kluwer.

Munévar, G. 1991. *Beyond Reason: Essays on the Philosophy of Paul K. Feyerabend*. London: Kluwer.

Munévar, G. 2000. "A Réhabilitation of Paul Feyerbend". In *The Worst Enemy of Science?: Essays in Memory of Paul Feyerabend*, J. Preston, G. Munévar, and D. Lamb (eds.), Oxford: Oxford University Press, 58–79.

Næss, A. 1975. "Why Not Science for Anarchists too? A Reply to Feyerabend." *Inquiry* 18: 183–194.

Naess, A. 1991. "Paul Feyerabend – a Green Hero?" In *Beyond Reason: Essays on the Philosophy of Paul K. Feyerabend*, G. Munévar (ed.), London: Kluwer, 403–416.

Nagel, E. 1977. "Against Method Review." *The American Political Science Review* 17: 1132–1134.

Nietzsche, F. 1968. *Twilight of the Idols*. R. Hollingdale (trans.). London: Penguin.

O'Flaherty, W. 1979. "Sacred Cows and Profane Mares in Indian Mythology." *History of Religions 19*(1): 1–26.

O'Neill, O. 1992. "Vindicating Reason." In *The Cambridge Companion to Kant*, P. Guyer, J. Nelson (eds.), Cambridge: Cambridge University Press, 280–308.

Oberheim, E. 2005. "On the Historical Origins of the Contemporary Notion of Incommensurability: Paul Feyerabend's Assault on Conceptual Conservativism." *Studies in History and Philosophy of Science* 36: 363–390.

Oberheim, E. 2006. *Feyerabend's Philosophy*. Berlin: Walter de Gruyter.

Oberheim, E. and P. Hoyningen-Huene. 2000. "Feyerabend's Early Philosophy." *Studies in History and Philosophy of Science* 31(2): 363–375.

Olin, M. 1992. *Forms of Representation in Alois Riegl's Theory of Art*. University Park: Pennsylvania State University Press.

Ottinger, G. 2010. "Buckets of Resistance: Standards and the Effectiveness of Citizen Science." *Science, Technology, & Human Values* 35(2): 244–270.

Overdevest, C. and B. Mayer. 2008. "Harnessing the Power of Information through Community Monitoring: Insights from Social Science." *Texas Law Review* (86)7: 1493–1526.

Papineau, D. 2014. "The Poverty of Conceptual Analysis." In *Philosophical Methodology: The Armchair or the Labouratory*, R. Batterman (ed.), Abingdon: Routledge, 166–194.

Patton, L. 2012. "Experiment and Theory Building." *Synthese* 184(3): 235–246.

Patton, P., Overgaard, N. and Barseghyan, H. 2017. "Reformulating the Second Law." *Scientonomy* 1: 29–39.

Peels, R. 2018. "A Conceptual Map of Scientism." In *Scientism: Problems and Prospects*, J. de Ridder, R. Peels, and R. van Woudenberg (eds.), Oxford: Oxford University Press, 28–56.

Penrose, R. 1979. "Singularities and Time-Asymmetry." In *General Relativity: An Einstein Centenary Survey*, S. Hawking and W. Israel (eds.), Cambridge: Cambridge University Press, 581–638.

Pielke Jr., R. 2007. *The Honest Broker: Making Sense of Science in Policy and Politics*. Cambridge: Cambridge University Press.

Pinto de Oliveira, J. 2017. "Image of Art: The First Manuscript of Structure." *Perspectives on Science* 25(6): 746–765.

Place, U. T. 1970. "Is Consciousness a Brain Process?" In *The Mind-Brain Identity Theory*, D. Armstrong (ed.), Palgrave: London, 42–51.

Popper, K. 1945. *The Open Society and Its Enemies*. London: Routledge & Kegan Paul.

Popper, K. 1949. "Naturgesetze und theoretische Systeme." In *Gesetz und Wirklichkeit. Internationale Hochschulwochen des Österreichischen College Alpbach, Tirol, 21. 8.-9.9.1948*, S. Moser (ed.), II. Innsbruck: Tyrolia-Verlag, 43–60.

Popper, K. 1956. "Three Views Concerning Human Knowledge." In *Contemporary British Philosophy*, H. Lewis (ed.), 357–88. Third Series. London: Allen & Unwin.

Popper, K. 1959. *The Logic of Scientific Discovery*. London: Hutchinson & Co.

Popper, K. 1963. *Conjectures and Refutations*. New York: Routledge.

Popper, K. 1983. *Realism and the Aim of Science.* London: Hutchinson & Co.

Prainsack, B. 2011. "Voting with their Mice: Personal Genome Testing and the 'Participatory Turn' in Disease Research." *Accountability in Research Policies and Quality Assurance* 18(3): 132–147.

Preston, J. 1997a. *Feyerabend: Philosophy, Science and Society.* Polity Press: Cambridge.

Preston, J. 1997b. "Feyerabend's Retreat from Realism." *Philosophy of Science* 64: S421–31.

Preston, J. 2000. "Science as Supermarket: 'Postmodern' Themes in Paul Feyerabend's Later Philosophy of Science." In *The Worst Enemy of Science? Essays in memory of Paul Feyerabend,* J. Preston, G. Munévar and D. Lamb (eds.), New York and Oxford: Oxford University Press, 80–101.

Preston, J. 2016. "The Rise of Western Rationalism: Paul Feyerabend's Story." *Studies in History and Philosophy of Science* 57: 79–86.

Preston, J. 2017. "Paul Feyerabend and the Debate Over the Philosophy of Science." *Oxford University Press Blogs.* https://blog.oup.com/2017/03/paul-feyerabend-philosophy-science/. Last date accessed: November 18, 2020.

Priest, G. 2005. *Doubt Truth to be a Liar.* Oxford: Oxford University Press.

Putnam, H. 1967. "Psychological Predicates." In *Art, Mind, and Religion,* In D. Armstrong (ed.), Pittsburgh: University of Pittsburgh Press, 37–48.

Putnam, H. 1987. *The Many Faces of Realism.* La Salle: Open Court.

Putnam, H. 1992. *Realism with a Human Face,* ed. by James Conant. Cambridge, MA: Harvard University Press.

Quine, W. 1951. "On Carnap's Views on Ontology." *Philosophical Studies,* 2(5): 65–72.

Quine, W. 1992. *Pursuit of Truth,* Revised edition. Cambridge, MA: Harvard University Press.

Rampley, M. 2013. *The Vienna School of Art History.* University Park: Pennsylvania State University Press.

Ravetz, J. 1971. *Scientific Knowledge and its Social Problems.* Oxford: Oxford University Press.

Rawleigh, W. 2018. "The Status of Questions in the Ontology of Scientific Change." *Scientonomy* 2: 1–12.

Reisch, G. 2005. *How the Cold War Transformed Philosophy of Science: To the Icy Slopes of Logic.* Cambridge: Cambridge University Press.

Riegl, A. 1893/1992. *Problems of Style: Foundations for a History of Ornament.* Evelyn Kain (trans.). Princeton: Princeton University Press.

Riegl, A. 1901/1985. *Late Roman Art Industry,* R. Winkes (trans.). Rome: Giorgio Bretschneider Editore.

Roe, S. 2009. "The Attenuated Ramblings of a Mad Man: Feyerabend's Anarchy Examined." *Polish Journal of Philosophy* 3(2): 67–85.

Rookmaaker, L. C. 1976. "An Early Engraving of the Black Rhinoceros (Diceros bicornis (L.)) Made by Jan Wandelaar." *Journal of the Linnean Society* 8: 87–90.

Rorty, R. 1965. "Mind-Body Identity, Privacy, and Categories." *The Review of Metaphysics,* 19(1): 24–54.

Rorty, R. 1982. "Contemporary Philosophy of Mind." *Synthese*, 53(2): 323–348.

Rudner, R. 1953. "The Scientist Qua Scientist Makes Value Judgments." *Philosophy of Science* 20(1): 1–6.

Russell, B. 1912. *The Problems of Philosophy*. Oxford: Oxford University Press.

Sankey, H. 1994. "Relativism and Epistemological Anarchism." *Cogito* 8(2): 158–164.

Sarewitz,. 2016. "Saving Science." *The New Atlantis* 49: 4–40.

Sbierski, J. 2018. "On the Proof of the C^0-Inextendibility of the Schwarzschild Spacetime." *Journal of Physics: Conference Series*, 968: 012012.

Scheffler, I. 1999. "A Plea for Pluralism." *Transactions of the Charles S. Peirce Society* 35: 425–36.

Schneider, N. 2009. "Form of Thought and Representational Gesture in Karl Popper and E.H Gombrich." *Human Affairs: A Postdisciplinary Journal for Humanities & Social Sciences* 19(3): 251–258.

Schofield, C. 1981. *Tychonic and Semi-Tychonic World Systems*. New York: Arno Press.

Sebastien, Z. 2016. "The Status of Normative Propositions in the Theory of Scientific Change." *Scientonomy* 1: 1–9.

Selinger, E. 2003. "Feyerabend's Democratic Critique of Expertise." *Critical Review* 15(3–4): 359–373.

Seymour, V. and Haklay, M. 2017. "Exploring Engagement Characteristics and Behaviours of Environmental Volunteers." *Citizen Science: Theory and Practice* 2(1): 1–13.

Shank, M. 2002. "Regiomontanus on Ptolemy, Physical Orbs, and Astronomical Fictionalism: Goldsteinian Themes in the 'Defense of Theon against George of Trebizond'." *Perspectives on Science* 10(2): 179–207.

Shapin, S. 1989. "The Invisible Technician." *American Scientist* 77(6): 554–563.

Shapiro, L. 2008. "How to Test for Multiple Realization." *Philosophy of Science* 75 (5): 514–525.

Shaw, J. 2017. "Was Feyerabend an Anarchist?: The Structure(s) of 'Anything Goes'." *Studies in History and Philosophy of Science Part A* 64: 11–21.

Shaw, J. 2018a. *A Pluralism Worth Having: Feyerabend's Well-Ordered Science*. Doctoral dissertation: University of Western Ontario.

Shaw, J. 2018b. "Feyerabend's Well-Ordered Science: How an Anarchist Distributes Funds." *Synthese* 1–31.

Shaw, J. 2020. "The Revolt Against Rationalism: Feyerabend's Critical Philosophy." *Studies in History and Philosophy of Science, Part A* 80: 110–122.

Shaw, J. and Barseghyan, H. 2019. "Problems and Prospects with the Scientonomic Workflow." *Scientonomy* 3: 1–14.

Sismondo, S. 2005. "Boundary Work and the Science Wars: James Robert Brown's Who Rules in Science?" *Episteme*, 1(3): 235–248.

Smaldino, P. and R. McElreath. 2016. "The Natural Selection of Bad Science." *Royal Society Open Science* 3(9).

Smith, P. 2004. *The Body of the Artisan*. Chicago: University of Chicago Press.

Smolin, L. 2006. *The Trouble With Physics*. Boston: Houghton Mifflin Harcourt.

Sokal, A. and Bricmont, J. 1998. *Fashionable Nonsense: Postmodern Intellectuals' Abuse of Science*. New York: Picador.

Solovey, M. 2013. *Shaky Foundations: The Politics-Patronage-Social Science Nexus in Cold War America*. New Brunswick: Rutgers University Press.

Sorgner, H. 2016. "Challenging Expertise: Paul Feyerabend vs. Harry Collins & Robert Evans on Democracy, Public Participation and Scientific Authority: Paul Feyerabend vs. Harry Collins & Robert Evans on Scientific Authority and Public Participation." *Studies in History and Philosophy of Science Part A* 57: 114–120.

Stadler, F. 2010. "Paul Feyerabend and the Forgotten "Third Vienna Circle"." In *Vertreibung, Transformation and Ruckkehr der Wissenschatfstheorie: Am Beispiel von Rudolf Carnap und Wolfgang Stegmuller*, F. Stadler (ed.). Mit einem Manuskript von Paul Feyerabend uber "Die Dogmen des Logischen Empirismus" aus dem Nachlass. Wien-Berlin-Munster: LIT Verlag, 169–187.

Stadler, F. and Fischer, K. (eds.). 2006. *Paul Feyerabend: Ein Philosoph aus Wien*, Vienna: Springer.

Staley, K. 1999. "Logic, Liberty, and Anarchy: Mill and Feyerabend on Scientific Method." *The Social Science Journal* 36(4): 603–14.

Stanford, P. 2006. *Exceeding our Grasp: Science, History, and the Problem of Unconceived Alternatives*. Oxford: Oxford University Press.

Steel, D. 2016. "Climate Change and Second-Order Uncertainty: Defending a Generalized, Normative, and Structural Argument from Inductive Risk." *Perspectives on Science* 24(6): 696–721.

Stegenga, J. 2018. *Medical Nihilism*. Oxford: Oxford University Press.

Stenmark, M. 2001. *Scientism: Science, Ethics, and Religion*. Aldershot, UK: Ashgate.

Storksdieck, M., Shirk, J., Cappadonna, J., Domroese, M., Göbel, C., Haklay, M., Miller-Rushing, A., Roetman, P., Sbrocchi, C and Vohland, K. 2016. "Associations for Citizen Science: Regional Knowledge, Global Collaboration." *Citizen Science: Theory and Practice* 1(2): 1–10.

Stove, D. 1982. *Popper and After: Four Modern Irrationalists*. Oxford: Pergamon Press.

Strawson, P. 1992. *Analysis and Metaphysics: An Introduction to Philosophy*. Oxford: Oxford University Press.

Stuart, M. forthcoming in 2020. "The Productive Anarchy of Scientific Imagination." *Philosophy of Science*.

Tambolo, L. 2014. "Pliability and Resistance: Feyerabendian Insights into Sophisticated Realism." *European Journal for Philosophy of Science* 4: 197–213.

Terry, S. and C. Boyd. 2001. "Researching the Biology of PXE: Partnering in the Process." *American Journal of Medical Genetics* 106: 117–184.

Theobald, E., Ettinger, E., Burgess, H., DeBay, L., Schmidt, N., Froehlich, H., Wagner C., HilleRisLambers, J., Tewksbury J., Harsch, M. and Parrish, J. 2015. "Global Change and Local Solutions: Tapping the Unrealized Potential of Citizen Science for Biodiversity Research." *Biological Conservation* 181: 236–244.

Theocharis, T. and M. Psimopoulos. 1987. "Where Science Has Gone Wrong." *Nature* 329: 595–598.

Thoren, V. 1990. *The Lord of Uraniborg: A Biography of Tycho Brahe*. Cambridge: Cambridge University Press.

Tibbetts, P. 1977. "Feyerabend's 'Against Method': The Case for Methodological Pluralism." *Philosophy of the Social Science* 7: 265–275.

Tibbetts, P. 1978. "A Response to Feyerabend on Science and Magic." *Philosophy of the Social Science* 8: 55–57

Torretti, R. 2000. "'Scientific Realism' and Scientific Practice." In *The Reality of the Unobservable*, E. Agazzi and M. Pauri (eds.), Dordrecht: Kluwer, 113–22.

Trumbull, D., Bonney, R., Bascom, D. and Cabral, A. 2000. "Thinking Scientifically During Participation in a Citizen-Science Project." *Science Education* 84: 265–275.

Trumbull, D., Bonney, R. and Grudens-Schuck, N. 2005. "Developing Materials to Promote Inquiry." *Science Education* 89: 879–900.

Tsou, J. Y. 2003. "Reconsidering Feyerabend's 'Anarchism'." *Perspectives on Science* 11(2): 208–235.

Tsueng, G., Nanis, S., Fouquier, J., Good, B. and Su, A. 2016. "Citizen Science for Mining the Biomedical Literature." *Citizen Science: Theory and Practice* 1 (2): 1–11.

United States Environmental Protection Agency. 2019. *Handbook for Citizen Science: Quality Assurance and Documentation*. Accessed online on October 15, 2019. www.epa.gov/sites/production/files/2019–03/documents/508_csqap phandbook_3_5_19_mmedits.pdf.

van Fraassen, B. 1980. *The Scientific Image*. Oxford: Clarendon Press.

van Fraassen, B. 2000. "Review of Paul Feyerabend, *Conquest of Abundance*." *Times Literary Supplement 5073: 23 June 2000* 5073: 10–11.

van Fraassen, B. 2002. *The Empirical Stance*, Princeton: Princeton University Press.

van Fraassen, B. 2004a. "Replies." *Philosophical Studies* 121: 171–192.

van Fraassen, B. 2004b. "Reply to Chakravartty, Jauernig, and McMullin," unpublished typescript. Accessed last on September 13, 2018, from www .princeton.edu/~fraassen/abstract/ReplyAPA-04.pdf

van Fraassen, B. 2011. "On Stance and Rationality." *Synthese* 178: 155–169.

van Strien, M. 2020. "Pluralism and Anarchism in Quantum Physics: Paul Feyerabend's Writings on Quantum Physics in Relation to his General Philosophy of Science." *Studies in History and Philosophy of Science Part A* 80: 72–81.

Vasari, G. 1550/1979. *Lives of the Artists*, G. Bull (trans.). Harmondsworth: Penguin.

Viola, T. 2012. "Pregmatism, Bistable Images and the Serpentine Line: A Chapter in the Prehistory of the Duck-Rabbit." In *Das Bildnerische Denken: Charles S. Peirce*, F. Engel, M. Queisner and T. Viola (eds.), Berlin: Akademie Verlag, 115–134.

von Neumann, J. 1932/1955. *Mathematical Foundations of Quantum Mechanics*, R. Beyer (trans.). Princeton: Princeton University Press.

Wald, R. 1984. *General Relativity*. Chicago: University of Chicago Press.

West, S. and Pateman, R. 2016. "Recruiting and Retaining Participants in Citizen Science: What Can Be Learned from the Volunteering Literature?" *Citizen Science: Theory and Practice* 1(2): 1–10.

Westfall, R. 1971. *The Construction of Modern Science: Mechanisms and Mechanics.* Cambridge: Cambridge University Press.

Westman, R. 1975. "The Melanchthon Circle, Rheticus, and the Wittenberg Interpretation of the Copernican Theory." *Isis*, 66 (2): 164–193.

Westman, R. 1986/2003. "The Copernicans and the Church." In *The Scientific Revolution*, M. Hellyer, (ed.), Oxford: Blackwell Publishing, 46–71.

Whitt, L. 1990. "Theory Pursuit: Between Discovery and Acceptance." In *Proceedings of the Biennial Meeting of the PSA.* Chicago: University of Chicago Press, 467–483.

Wikipedia. 2018. "Carl von Clausewitz." in *Wikipedia, The Free Encyclopedia,* retrieved 15:30, September 18, 2018, from https://en.wikipedia.org/wiki/Carl_von_Clausewitz

Wilderman, C. and Monismith, J. 2016. "Monitoring Marcellus: A Case Study of a Collaborative Volunteer Monitoring Project to Document the Impact of Unconventional Shale Gas Extraction on Small Streams." *Citizen Science: Theory and Practice* 1(1): 1–17.

Wittgenstein, L. 1922. *Tractatus Logico-Philosophicus.* C. Ogden (trans.), London: Routledge & Kegan Paul.

Wittgenstein, L. 1953. *Philosophical Investigations.* Oxford: Basil Blackwell.

Wittgenstein, L. 1980. *Culture and Value,* G. H. von Wright (ed.), Peter Winch (trans.). Oxford: Blackwell.

Wolff, R. 1970. *In Defence of Anarchism.* Berkeley: University of California Press.

Wood, C. 2009. "E.H. Gombrich's Art and Illusion: A Study in the Psychology of Pictorial Representation, 1960." *The Burlington Magazine* 151(1281): 836–839.

Wood, C., Sullivan, B., Iliff, M., Fink, D. and Kelling, S. 2011. "eBird: Engaging Birders in Science and Conservation." *PLoS Biology* 9(12): 1–5.

Worrall, J. 1978. "Against Too Much Method." *Erkenntnis* 13: 279–295.

Worrall, J. 1988. "Review: The Value of a Fixed Methodology." *The British Journal for the Philosophy of Science* 39: 263–275.

Worrall, J. 1989. "Fix It and Be Damned: A Reply to Laudan." *The British Journal for the Philosophy of Science* 40(3): 376–388.

Wray, K. B. 2015. "The Methodological defence of Realism Scrutinized." *Studies in History and Philosophy of Science Part A* 54: 74–79.

Wykstra, S. 1980. "Toward a Historical Meta-Method for Assessing Normative Methodologies: Rationality, Serendipity, and the Robinson Crusoe Fallacy" in *Proceedings of the Biennial Meeting of the PSA.* The University of Chicago Press, 211–222.

Wylie, A. 2014. "Community-Based Collaborative Archaeology." In *Philosophy of Social Science: A New Introduction*, N. Cartwright and E. Montuschi (eds.), Oxford: Oxford University Press, 68–82.

Wylie, A. 2015. "A Plurality of Pluralisms: Collaborative Practice in Archaeology." In *Objectivity in Science: New Perspectives from Science and*

Technology Studies, F. Padovani, A. Richardson, and J. Tsou (eds.), New York: Springer, 189–210.

Wynne, B. 1989. "Sheepfarming after Chernobyl: A Case Study in Communicating Scientific Information." *Environment: Science and Policy for Sustainable Development* 31(2): 10–39.

Zahar, E. 1982. "Review: Feyerabend on Observation and Empirical Content". *The British Journal for the Philosophy of Science* 33(4): 397–409.

Zivin, J., Azoulay, P. and Fons-Rosen, C. 2019. "Does Science Advance One Funeral at a Time?" *American Economic Review* 109(8): 2889–2920.

Zooniverse. 2019. Accessed online on August 3, 2019. www.zooniverse.org/

Index

257

Printed in the United States
by Baker & Taylor Publisher Services

Printed in the United States
by Baker & Taylor Publisher Services